The Ubiquitous Quasidisk

Mathematical
Surveys
and
Monographs

Volume 184

The Ubiquitous Quasidisk

Frederick W. Gehring
Kari Hag

With contributions by
Ole Jacob Broch

American Mathematical Society
Providence, Rhode Island

2010 *Mathematics Subject Classification.* Primary 30C62;
Secondary 30C20, 30C45, 30C65, 30F45.

For additional information and updates on this book, visit
www.ams.org/bookpages/surv-184

Library of Congress Cataloging-in-Publication Data

Gehring, Frederick W.
The ubiquitous quasidisk / Frederick W. Gehring, Kari Hag ; with contributions by Ole Jacob Broch.
pages cm. — (Mathematical surveys and monographs ; volume 184)
Includes bibliographical references and index.
ISBN 978-0-8218-9086-8 (alk. paper)
1. Quasiconformal mappings. 2. Geometric function theory. 3. Functions of a complex variable. I. Hag, Kari, 1941– II. Broch, Ole Jacob. III. Title.

QA360.G44 2012
515′.9—dc23

2012030309

∞ The paper used in this book is acid-free and falls within the guidelines
established to ensure permanence and durability.
Visit the AMS home page at http://www.ams.org/

10 9 8 7 6 5 4 3 2 1 17 16 15 14 13 12

To Lois

Contents

Preface

In August 1981 the first author gave a short course of six lectures on function theory at the NATO Advanced Study Institute in Montreal. The object was to point out the surprising connections which quasidisks—by definition the images of a disk under a quasiconformal mapping of the Riemann sphere—have with various branches of analysis and geometry.

The written account, *Characteristic properties of quasidisks* (97 triple-spaced pages), was published by the University Press of the University of Montreal in 1982 and became quite popular. Thus the notes were out of print after a few years. In the meantime the number of characterizing properties increased, and in the late 1990s we decided to write a book with the tentative title *The Ubiquitous Quasidisk* as an expanded version of the "Montreal notes".

This book will hopefully be an inspiration for graduate students in geometric function theory. More specifically, the book could be a candidate for the text of a semester-long second-year graduate course on selected topics in the field. The texts by Ahlfors [**7**] and by Lehto and Virtanen [**117**] on quasiconformal mappings provide valuable reference literature. A more recent account for additional material is the book by Astala, Iwaniec, and Martin [**16**].

Our mathematical descendant Ole Jacob Broch has been of invaluable help in writing up the manuscript. He was assisted by Geir Arne Hjelle, another former student, who transformed most of the hand drawn figures to computer pictures in a way that we think helps to preserve the spirit of the "Montreal notes". We would also like to thank Per Hag and Olli Martio who have read the manuscript and made valuable suggestions.

Frederick W. Gehring
Kari Hag
Ann Arbor/Trondheim
September 2011

Part 1

Properties of quasidisks

CHAPTER 1

Preliminaries

We collect here some definitions and properties of plane quasiconformal mappings. Two basic references for this material are the books by Ahlfors [**7**] and Lehto and Virtanen [**117**], to which we refer the reader for further details. A more recent title is the book by Astala, Iwaniec, and Martin [**16**].

In what follows $\mathbf{R}^2$ denotes the Euclidean plane with its usual identification with the complex plane $\mathbf{C}$. The one-point compactification $\overline{\mathbf{R}}^2 = \mathbf{R}^2 \cup \{\infty\}$ is equipped with the chordal metric

$$\mathrm{ch}(z,w) = \frac{2|z-w|}{\sqrt{|z|^2+1}\,\sqrt{|w|^2+1}},$$

where we employ the usual conventions regarding ∞.

Let D and D' be subdomains of $\overline{\mathbf{R}}^2$. We will assume, unless stated otherwise, that $\mathrm{card}(\overline{\mathbf{R}}^2 \setminus D) \geq 2$. The exterior of D is denoted by $D^* = \overline{\mathbf{R}}^2 \setminus \overline{D}$. Let $\mathbf{B}(z,r)$ be the open Euclidean disk with center $z \in \mathbf{R}^2$ and radius r, and let $\mathbf{B}$ be the unit disk $\mathbf{B}(0,1)$. Finally, $\mathbf{H}$ will denote the upper or right half-planes

$$\{z = x + iy : y > 0\} \qquad \text{or} \qquad \{z = x + iy : x > 0\}.$$

1.1. Quasiconformal mappings

There are several different ways to view a quasiconformal mapping. Perhaps the most geometrically intuitive is in terms of the linear dilatation of a homeomorphism.

Suppose that $f : D \to D'$ is a homeomorphism. For $z \in D \setminus \{\infty, f^{-1}(\infty)\}$ and $0 < r < \mathrm{dist}(z, \partial D)$ we let

$$\begin{aligned} l_f(z,r) &= \min_{|z-w|=r} |f(z) - f(w)|, \\ L_f(z,r) &= \max_{|z-w|=r} |f(z) - f(w)| \end{aligned} \tag{1.1.1}$$

and call

$$H_f(z) = \limsup_{r \to 0} \frac{L_f(z,r)}{l_f(z,r)}$$

the *linear dilatation* of f at z. See Figure 1.1.

Recall that a homeomorphism in $\mathbf{R}^2$ is either sense-preserving or sense-reversing [**117**]. Menchoff showed in 1937 [**129**] that if $D, D' \subset \mathbf{R}^2$, a sense-preserving homeomorphism $f : D \to D'$ is analytic, and hence conformal, whenever

$$H_f(z) = 1 \tag{1.1.2}$$

for all but a countable set of $z \in D$.

The following definition for quasiconformality is a natural counterpart of Menchoff's theorem (Gehring [**47**]).

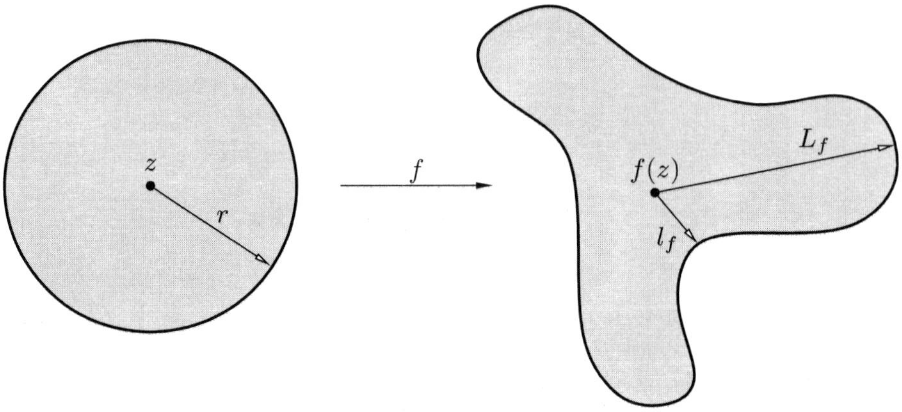

FIGURE 1.1

DEFINITION 1.1.3. A homeomorphism $f : D \to D'$ is K-*quasiconformal* where $1 \le K < \infty$ if $H_f(z) < \infty$ for every $z \in D \setminus \{\infty, f^{-1}(\infty)\}$ and

$$H_f(z) \le K$$

almost everywhere in D.

The inequality in the above definition can be weakened significantly to yield the same class of mappings. Letting

$$h_f(z) = \liminf_{r \to 0} \frac{L_f(z,r)}{l_f(z,r)},$$

where l_f and L_f are as in (1.1.1), Heinonen and Koskela [**81**] and Kallunki and Koskela [**96**] obtained the following surprising result.

THEOREM 1.1.4. *A homeomorphism $f : D \to D'$ is K-quasiconformal, where $1 \le K < \infty$, if $H_f(z) < \infty$ for every $z \in D \setminus \{\infty, f^{-1}(\infty)\}$ and*

$$h_f(z) \le K$$

almost everywhere in D.

The next results, which can be found in Lehto-Virtanen [**117**], identify mappings which are 1-quasiconformal or which are the composition and inverses of quasiconformal mappings.

THEOREM 1.1.5. *A homeomorphism $f : D \to D'$ is 1-quasiconformal if and only if f or its complex conjugate $\bar{f}$ is a conformal mapping, i.e., analytic in $D \setminus \{\infty, f^{-1}(\infty)\}$.*

THEOREM 1.1.6. *If $f : D \to D'$ is K_1-quasiconformal and $g : D' \to D''$ is K_2-quasiconformal, then $g \circ f : D \to D''$ is $K_1 K_2$-quasiconformal. The inverse of a K-quasiconformal mapping is K-quasiconformal.*

Menchoff's theorem asserts that a sense-preserving homeomorphism f of D is a conformal mapping if, except at a countable set of points $z \in D$, f maps infinitesimal circles about z onto infinitesimal circles about $f(z)$. Theorems 1.1.4 and 1.1.5 extend this result by first replacing the countable exceptional set where

(1.1.2) was not required to hold by a set of *measure zero* and then by requiring that f preserves only a *sequence* of infinitesimal circles about the remaining points $z \in D$.

DEFINITION 1.1.7. A real-valued function u is *absolutely continuous on lines*, or *ACL*, in a domain D if for each rectangle $[a, b] \times [c, d] \subset D$,

1° $u(x + i\,y)$ is absolutely continuous in x for almost all $y \in [c, d]$,
2° $u(x + i\,y)$ is absolutely continuous in y for almost all $x \in [a, b]$.

A complex-valued function f is ACL in D if its real and imaginary parts are ACL in D.

If a homeomorphism f is ACL in D, then a measure theoretic argument shows that f has finite partial derivatives a.e. in D and hence, in fact, a differential a.e. in D by Gehring-Lehto [**63**].

A quasiconformal mapping can then be described in terms of its analytic properties as follows. See e.g. Lehto-Virtanen [**117**].

THEOREM 1.1.8. *A homeomorphism $f : D \to D'$ is K-quasiconformal if and only if f is ACL in D and*

$$\max_{\alpha} |\partial_\alpha f(z)|^2 \le K\,|J_f(z)| \tag{1.1.9}$$

almost everywhere in D. Here $\partial_\alpha f(z)$ denotes the derivative of f at z in the direction α and $J_f(z)$ denotes the Jacobian of f at z. Moreover, if f is quasiconformal, we have that $J_f(z) \neq 0$ a.e. in D and that it satisfies Lusin's property (N)*, i.e. $m(f(E)) = 0$ whenever $m(E) = 0$ for the planar Lebesgue measure m.*

If $f : D \to D'$ is K-quasiconformal, then inequality (1.1.9) can also be written

$$\max_{\alpha} |\partial_\alpha\, f(z)| \le K \min_{\alpha} |\partial_\alpha\, f(z)|.$$

If we assume also that f is sense-preserving, then

$$\max_{\alpha} |\partial_\alpha\, f| = |f_z| + |f_{\bar z}|,$$
$$\min_{\alpha} |\partial_\alpha\, f| = |f_z| - |f_{\bar z}|,$$

where f_z and $f_{\bar z}$ are the *complex derivatives*

$$f_z = \frac{1}{2}(f_x - i f_y) \qquad \text{and} \qquad f_{\bar z} = \frac{1}{2}(f_x + i f_y).$$

In this case (1.1.9) takes the form

$$|f_{\bar z}| \le \frac{K-1}{K+1}\,|f_z|. \tag{1.1.10}$$

Then since

$$|f_z|^2 - |f_{\bar z}|^2 = J_f > 0$$

a.e. in D, we may also consider the quotient

$$\mu_f = \frac{f_{\bar z}}{f_z}.$$

The function $\mu_f(z)$ is the *complex dilatation* of f at z. It satisfies the relations

$$\frac{1 + |\mu_f(z)|}{1 - |\mu_f(z)|} = H_f(z) \qquad \text{and} \qquad |\mu_f(z)| \le \frac{K-1}{K+1}$$

a.e. in D. Hence $\mu_f = 0$ a.e. in D if and only if f is conformal.

If $f : D \to D'$ and $g : D' \to D''$ are both sense-preserving and quasiconformal, then $\mu_{g \circ f} = \mu_f$ a.e. in D if and only if g is conformal.

It is possible to prescribe the complex dilatation $\mu_f(z)$, and hence the linear dilatation $H_f(z)$, at almost every point z of a domain D. This result, known as the *measurable Riemann mapping theorem*, has turned out to be a powerful tool in complex analysis. See Ahlfors-Bers [**9**], Lehto-Virtanen [**117**], Morrey [**133**], and Bojarski [**25**].

THEOREM 1.1.11. *If μ is measurable with*

$$\|\mu\|_{L^\infty} = \operatorname*{ess\,sup}_D |\mu(z)| < 1,$$

then there exists a sense-preserving quasiconformal mapping f of D with $\mu_f = \mu$ a.e. in D. Moreover f is unique up to post composition with a conformal map.

1.2. Modulus of a curve family

The conditions for quasiconformality in Definition 1.1.3 and Theorem 1.1.8 involve the local behavior of a homeomorphism. We need a way to *integrate* the inequality in Theorem 1.1.8 in order to derive global properties of the mapping. When $K = 1$, f or its complex conjugate $\bar{f}$ is conformal and the Cauchy integral formula is available. The tool most often used to replace the Cauchy formula when $K > 1$ is the method of extremal length, first formulated by Ahlfors and Beurling in [**23**].

Suppose that Γ is a family of curves in $\overline{\mathbf{R}}^2$. We say that ρ is an *admissible density* for Γ, or is in $\mathrm{adm}(\Gamma)$, if ρ is nonnegative and Borel measurable in $\mathbf{R}^2$ and if

$$\int_\gamma \rho(z)|dz| \geq 1$$

for each locally rectifiable $\gamma \in \Gamma$. The *modulus* and *extremal length* of the family Γ are then given, respectively, by

$$\mathrm{mod}(\Gamma) = \inf_\rho \int_{\mathbf{R}^2} \rho(z)^2\, dm \qquad \text{and} \qquad \lambda(\Gamma) = \frac{1}{\mathrm{mod}(\Gamma)},$$

where the infimum is taken over $\rho \in \mathrm{adm}(\Gamma)$.

THEOREM 1.2.1. *If $f : D \to D'$ is conformal and if Γ is a family of curves in D, then*

$$\mathrm{mod}(f(\Gamma)) = \mathrm{mod}(\Gamma).$$

PROOF. We consider the case where $D, D' \subset \mathbf{R}^2$. For each $\rho' \in \mathrm{adm}(f(\Gamma))$ let

$$\rho(z) = \begin{cases} \rho'(f(z))|f'(z)| & \text{if } z \in D, \\ 0 & \text{if } z \in \mathbf{R}^2 \setminus D. \end{cases}$$

Then ρ is nonnegative and Borel measurable in $\mathbf{R}^2$. If γ is locally rectifiable, then $f(\gamma) \in f(\Gamma)$ is locally rectifiable and

$$\int_\gamma \rho(z)|dz| = \int_\gamma \rho'(f(z))|f'(z)||dz| = \int_{f(\gamma)} \rho'(w)|dw| \geq 1.$$

Thus $\rho \in \mathrm{adm}(\Gamma)$,

$$\begin{aligned}\mathrm{mod}(\Gamma) &\leq \int_{\mathbf{R}^2} \rho(z)^2 dm = \int_D \rho'(f(z))^2 |f'(z)|^2 dm \\ &= \int_{D'} \rho'(w)^2 dm \leq \int_{\mathbf{R}^2} \rho'(w)^2 dm,\end{aligned}$$

whence

$$\mathrm{mod}(\Gamma) \leq \inf_{\rho'} \int_{\mathbf{R}^2} \rho'(w)^2 dm = \mathrm{mod}(f(\Gamma)).$$

Now take the infimum over all such ρ.

Finally we obtain

$$\mathrm{mod}(\Gamma) = \mathrm{mod}(f(\Gamma))$$

by applying the above argument to f^{-1}. □

If the curves $\gamma \in \Gamma$ are disjoint arcs, we may think of them as homogeneous electric wires. Then the modulus $\mathrm{mod}(\Gamma)$ is a conformally invariant electrical transconductance for the family of wires γ and the extremal length $\lambda(\Gamma)$ is the total electrical resistance of the system. In particular, $\mathrm{mod}(\Gamma)$ is big if the curves $\gamma \in \Gamma$ are short and plentiful and small if the curves γ are long or scarce.

The following properties show that $\mathrm{mod}(\Gamma)$ is also an outer measure on the curve families Γ in $\overline{\mathbf{R}}^2$:

1° $\mathrm{mod}(\emptyset) = 0$.
2° $\mathrm{mod}(\Gamma_1) \leq \mathrm{mod}(\Gamma_2)$ if $\Gamma_1 \subset \Gamma_2$.
3° $\mathrm{mod}(\bigcup_j \Gamma_j) \leq \sum_j \mathrm{mod}(\Gamma_j)$.

Finally the conformal invariant $\mathrm{mod}(\Gamma)$ yields a third characterization for quasiconformal mappings.

THEOREM 1.2.2 (Ahlfors [**7**]). *A homeomorphism $f : D \to D'$ is K-quasiconformal if and only if*

$$\frac{1}{K}\,\mathrm{mod}(\Gamma) \leq \mathrm{mod}(f(\Gamma)) \leq K\,\mathrm{mod}(\Gamma)$$

for each family Γ of curves in D.

1.3. Modulus estimates

Estimates for the moduli of various curve families are useful tools for studying geometric properties of conformal and quasiconformal mappings. We derive here three simple modulus estimates and a distortion theorem for quasiconformal mappings of the plane which we will need later.

LEMMA 1.3.1. *Suppose that $R = R(0, a, a+i, i)$ is the rectangle with vertices at $0, a, a+i, i$ where $a > 0$ and suppose that Γ is the family of curves which join the horizontal sides of ∂R in R. Then*

$$\mathrm{mod}(\Gamma) = a.$$

PROOF. The segment $\gamma = \{z : x + iy : 0 < y < 1\}$ is in Γ for $0 < x < a$. Hence if $\rho \in \mathrm{adm}(\Gamma)$, then by the Cauchy-Schwarz inequality,

$$1 \leq \left(\int_0^1 \rho(x+iy)\,dy\right)^2 \leq \int_0^1 \rho(x+iy)^2\,dy$$

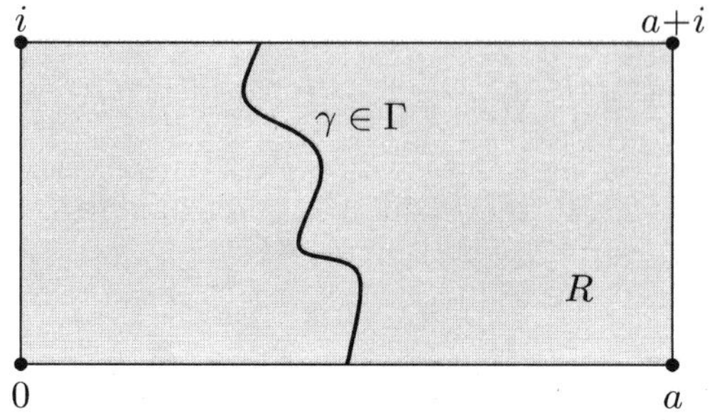

FIGURE 1.2

for $0 < x < a$. Thus

$$\int_{\mathbf{R}^2} \rho(z)^2 \, dm \geq \int_0^a \left(\int_0^1 \rho(x + i\, y)^2 \, dy \right) dx \geq a$$

and

$$\operatorname{mod}(\Gamma) = \inf_\rho \int_{\mathbf{R}^2} \rho(z)^2 \, dm \geq a.$$

Next the function

$$\rho(z) = \begin{cases} 1 & \text{if } z \in R, \\ 0 & \text{otherwise} \end{cases}$$

is in $\operatorname{adm}(\Gamma)$ and

$$\int_{\mathbf{R}^2} \rho(z)^2 \, dm = a,$$

completing the proof for Lemma 1.3.1. □

LEMMA 1.3.2. *If Γ is a family of curves and if for each t with $a < t < b$ the circle $\{z : |z| = t\}$ contains a curve $\gamma \in \Gamma$, then*

$$\operatorname{mod}(\Gamma) \geq \frac{1}{2\pi} \log \frac{b}{a}.$$

PROOF. See Figure 1.3. If $\rho \in \operatorname{adm}(\Gamma)$, then

$$1 \leq \left(\int_\gamma \rho(z) \, |dz| \right)^2 \leq \left(\int_0^{2\pi} \rho(t\, e^{i\,\theta}) \, t \, d\theta \right)^2 \leq 2\pi t \int_0^{2\pi} \rho(t\, e^{i\,\theta})^2 \, t \, d\theta,$$

whence

$$\frac{1}{2\pi} \log \frac{b}{a} = \int_a^b \frac{1}{2\pi t} \, dt \leq \int_a^b \left(\int_0^{2\pi} \rho(t\, e^{i\,\theta})^2 t \, d\theta \right) dt \leq \int_{\mathbf{R}^2} \rho(z)^2 \, dm.$$

Now take the infimum over all such ρ. □

LEMMA 1.3.3. *If Γ is a family of curves which join continua C_1 and C_2 where*

$$\operatorname{dist}(C_1, C_2) \geq a > 0, \qquad \operatorname{diam}(C_1) \leq b,$$

then

$$\operatorname{mod}(\Gamma) \leq \pi \left(\frac{b}{a} + 1 \right)^2.$$

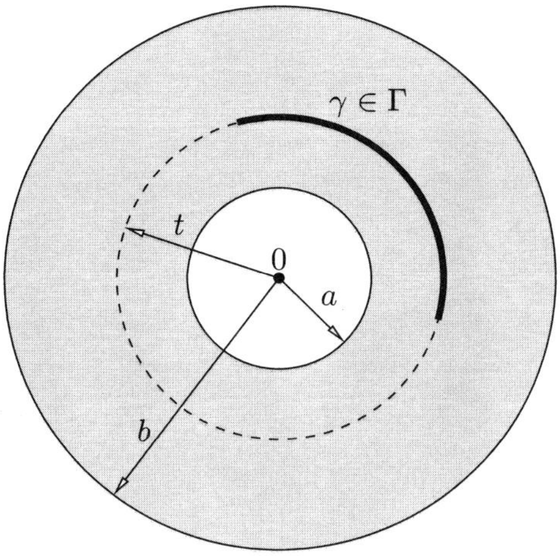

FIGURE 1.3

PROOF. Choose $z_1 \in C_1$ and $z_2 \in C_2$ so that $|z_1 - z_2| = \operatorname{dist}(C_1, C_2)$ and set

$$\rho(z) = \begin{cases} 1/a & \text{if } z \in \mathbf{B}(z_1, a+b), \\ 0 & \text{otherwise.} \end{cases}$$

Then each $\gamma \in \Gamma$ either joins C_1 to C_2 in $\mathbf{B}(z_1, a+b)$ or joins $\partial\mathbf{B}(z_1, b)$ to $\partial\mathbf{B}(z_1, a+b)$. In either case γ contains a subarc of length at least a which lies in $\mathbf{B}(z_1, a+b)$. Thus $\rho \in \operatorname{adm}(\Gamma)$ and

$$\operatorname{mod}(\Gamma) \leq \int_{\mathbf{R}^2} \rho(z)^2 \, dm = \pi\Big(\frac{b}{a} + 1\Big)^2.$$

□

We now apply the modulus estimate established above in Lemma 1.3.2 to prove an elementary distortion theorem for quasiconformal mappings which we will need in what follows.

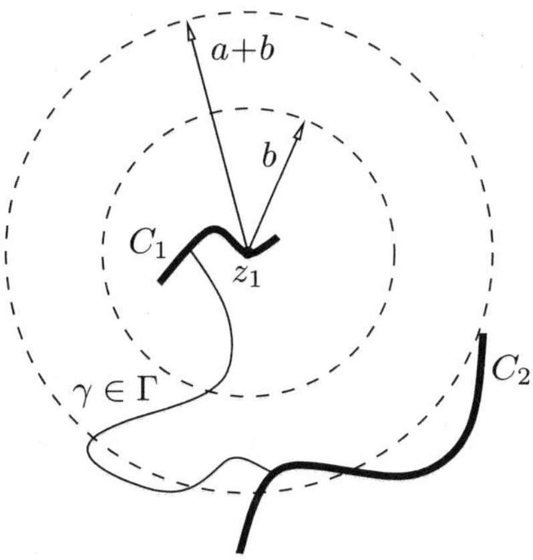

FIGURE 1.4

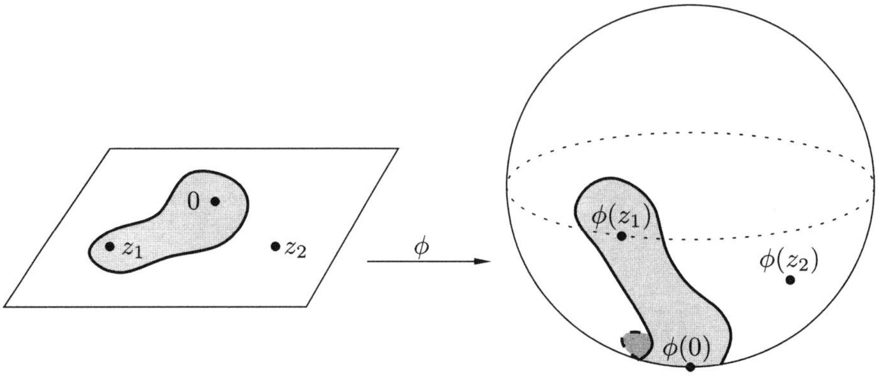

FIGURE 1.5

THEOREM 1.3.4. *If $f : \mathbf{R}^2 \to \mathbf{R}^2$ is K-quasiconformal and if*

$$|z_2 - z_0| \leq |z_1 - z_0|,$$

then

$$|f(z_2) - f(z_0)| \leq c\,|f(z_1) - f(z_0)| \tag{1.3.5}$$

where $c = e^{8K}$.

PROOF. By means of preliminary similarity transformations, we may assume that $z_0 = f(z_0) = 0$ and that $|z_1| = 1$, whence $|z_2| \leq 1$. We may also assume that $|f(z_1)| < |f(z_2)|$ since otherwise there is nothing to prove.

Let Γ' be the family of circles $\{w : |w| = t\}$ where $|f(z_1)| < t < |f(z_2)|$. Then

$$\frac{1}{2\pi} \log \frac{|f(z_2)|}{|f(z_1)|} \leq \operatorname{mod}(\Gamma')$$

by Lemma 1.3.2.

To estimate the modulus of $\Gamma = f^{-1}(\Gamma')$, let ϕ denote the stereographic projection of $\overline{\mathbf{R}}^2$ onto the Riemann sphere $\mathbf{S}^2 = \{x \in \mathbf{R}^3 : |x| = 1\}$. If $\gamma \in \Gamma = f^{-1}(\Gamma')$, then γ separates the points 0 and z_1 from ∞ and z_2; hence $\phi(\gamma)$ is a closed curve on $\mathbf{S}^2$ which separates the points $\phi(0)$ and $\phi(z_1)$ from $\phi(\infty)$ and $\phi(z_2)$. Since each arc on $\mathbf{S}^2$ which joins $\phi(0)$ to $\phi(z_1)$ or $\phi(\infty)$ to $\phi(z_2)$ has length at least $\pi/2$,

$$\int_\gamma \frac{2}{1 + |z|^2}\,|dz| = \operatorname{length}(\phi(\gamma)) \geq \pi$$

and hence the density

$$\rho(z) = \frac{1}{\pi}\,\frac{2}{1 + |z|^2}$$

is admissible for Γ. Thus

$$\operatorname{mod}(\Gamma) \leq \int_{\mathbf{R}^2} \rho(z)^2\,dm = \frac{1}{\pi^2} \int_{\mathbf{R}^2} \frac{4}{(1 + |z|^2)^2}\,dm = \frac{4}{\pi}$$

and we obtain

$$\frac{1}{2\pi} \log \frac{|f(z_2)|}{|f(z_1)|} \leq \operatorname{mod}(\Gamma') \leq K \operatorname{mod}(\Gamma) \leq \frac{4K}{\pi}$$

from which (1.3.5) follows. $\square$

A more detailed reasoning yields the following sharp estimate for the constant c in (1.3.5), namely $c = \lambda(K)$ where

$$\lambda(K) = \left(\frac{1}{4}\,e^{\pi K/2} - e^{-\pi K/2}\right)^2 + \delta(K), \qquad 0 < \delta(K) < e^{-\pi K}. \tag{1.3.6}$$

See Anderson-Vamanamurthy-Vuorinen [**11**] and Lehto-Virtanen-Väisälä [**118**].

COROLLARY 1.3.7. *If $f : \mathbf{R}^2 \to \mathbf{R}^2$ is K-quasiconformal and if*

$$|z_2 - z_0| \le 2^k\,|z_1 - z_0| \tag{1.3.8}$$

where k is an integer, $k \ge 0$, then

$$|f(z_2) - f(z_0)| \le c\,(c+1)^k\,|f(z_1) - f(z_0)| \tag{1.3.9}$$

where $c = e^{8K}$.

PROOF. By Theorem 1.3.4, (1.3.8) implies (1.3.9) when $k = 0$. Suppose this implication is true for some $k \ge 0$ and set $z = \frac{1}{2}(z_2 + z_0)$. Then

$$|z_2 - z| = |z - z_0| \le 2^k\,|z_1 - z_0|$$

and

$$|f(z_2) - f(z)| \le c\,|f(z) - f(z_0)|$$

again by Theorem 1.3.4. Since

$$|f(z) - f(z_0)| \le c(c+1)^k\,|f(z_1) - f(z_0)|$$

by hypothesis, we obtain

$$\begin{aligned} |f(z_2) - f(z_0)| &\le |f(z_2) - f(z)| + |f(z) - f(z_0)| \\ &\le (c+1)\,|f(z) - f(z_0)| \\ &\le c(c+1)^{k+1}\,|f(z_1) - f(z_0)|. \end{aligned}$$

Thus (1.3.8) implies (1.3.9) for $k+1$ and hence for all k by induction. □

Theorem 1.3.4 and its corollary are also consequences of the following general result (Gehring-Hag [**57**]), the proof of which is less elementary and depends on theorems due to Teichmüller [**157**] and Agard [**1**].

THEOREM 1.3.10. *If $f : \mathbf{R}^2 \to \mathbf{R}^2$ is K-quasiconformal, then*

$$\frac{|f(z_2) - f(z_0)|}{|f(z_1) - f(z_0)|} + 1 \le 16^{K-1}\left(\frac{|z_2 - z_0|}{|z_1 - z_0|} + 1\right)^K$$

for $z_0, z_1, z_2 \in \mathbf{R}^2$. The coefficient 16^{K-1} cannot be replaced by any smaller constant.

The property in Theorem 1.3.10 is called *quasisymmetry* (Heinonen [**80**], Astala-Iwaniec-Martin [**16**]).

We conclude by listing two properties of quasiconformal mappings that we will need in what follows. See, for example, Lehto-Virtanen [**117**].

THEOREM 1.3.11. *If $f : D \to D'$ is quasiconformal and if D and D' are Jordan domains, then f has a homeomorphic extension which maps $\overline{D}$ onto $\overline{D'}$.*

THEOREM 1.3.12. *Suppose that $E \subset D$ is closed and contained in a countable union of rectifiable curves. If $f : D \to D'$ is a homeomorphism which is K-quasiconformal in each component of $D \setminus E$, then f is K-quasiconformal in D.*

1.4. Quasidisks

We come now to the principal object of study in this book.

DEFINITION 1.4.1. A domain D is a K-*quasidisk* if it is the image of a Euclidean disk or half-plane under a K-quasiconformal self-mapping of $\overline{\mathbf{R}}^2$. D is a *quasidisk* if it is a K-quasidisk for some K.

We present next three Jordan domains that we will use in what follows to illustrate various properties of quasidisks. The first of these is an angular sector.

EXAMPLE 1.4.2. For $0 < \alpha < 2\pi$ let $\mathbf{S}(\alpha)$ denote the angular sector

$$\mathbf{S}(\alpha) = \{z = r\,e^{i\,\theta} : 0 < r < \infty,\ |\theta| < \frac{\alpha}{2}\}.$$

Then $\mathbf{S}(\alpha)$ is a K-quasidisk where

$$K = \max\left(\sqrt{\frac{2\,\pi - \alpha}{\alpha}}, \sqrt{\frac{\alpha}{2\,\pi - \alpha}}\right). \tag{1.4.3}$$

The bound in (1.4.3) is sharp.

To prove this, let

$$f(r\,e^{i\,\theta}) = r^p\,e^{i\,\phi(\theta)}$$

for $0 < r < \infty$ and $|\theta| \le \pi$ where

$$p = \frac{\pi}{\sqrt{(2\,\pi - \alpha)\,\alpha}}$$

and

$$\phi(\theta) = \begin{cases} \dfrac{\pi\,\theta}{\alpha} & \text{if } \ 0 \le \theta \le \dfrac{\alpha}{2}, \\ \pi - \dfrac{\pi\,(\pi - \theta)}{2\,\pi - \alpha} & \text{if } \ \dfrac{\alpha}{2} \le \theta \le \pi, \\ -\phi(-\theta) & \text{if } \ -\pi \le \theta \le 0. \end{cases}$$

An elementary calculation shows that f is K-quasiconformal, where K is as in (1.4.3), and that f maps $\mathbf{S}(\alpha)$ onto the right half-plane $\mathbf{S}(\pi)$.

To show that the bound in (1.4.3) is best possible, suppose that f is a K-quasiconformal mapping of $\overline{\mathbf{R}}^2$ which maps $\mathbf{S}(\alpha)$ onto the right half-plane $\mathbf{S}(\pi)$ and let $h = f^{-1} \circ g \circ f$ where g denotes reflection in the imaginary axis. Then h is a K^2-quasiconformal mapping of $\overline{\mathbf{R}}^2$ which maps $\mathbf{S}(\alpha)$ onto its exterior $\mathbf{S}^*(\alpha)$.

Next fix $0 < a < b < \infty$ and let Γ denote the family of arcs which join the circles $\{z : |z| = a\}$ and $\{z : |z| = b\}$ in

$$\{z : a \le |z| \le b,\ |\arg(z)| < \alpha/2\}.$$

Then it is not difficult to check that

$$\operatorname{mod}(\Gamma) = \frac{\alpha}{\log(b/a)}.$$

Similarly,

$$\operatorname{mod}(\Gamma') = \frac{2\,\pi - \alpha}{\log(b/a) + 2\log(c)}$$

where Γ' is the family of arcs which join $\{z : |z| = a/c\}$ and $\{z : |z| = bc\}$ in

$$\{z : a/c \le |z| \le bc,\ \alpha/2 < |\arg(z)| \le \pi\}.$$

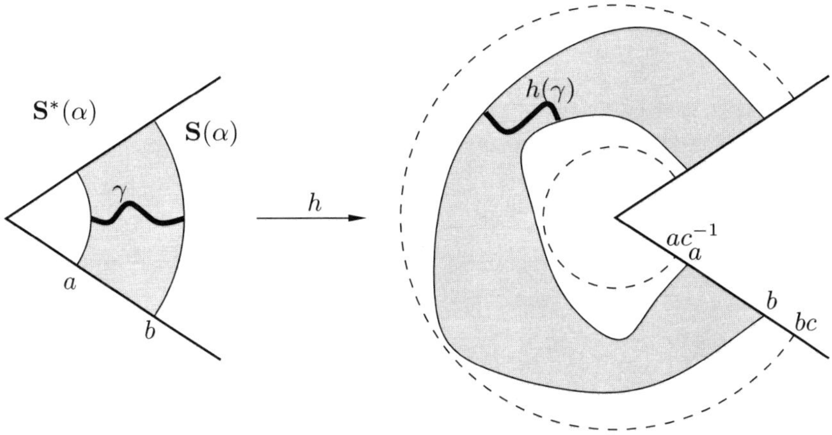

FIGURE 1.6

Hence if $c = 8\,e^{K^2}$, then Theorem 1.3.4 implies that for each arc $\gamma' \subset \Gamma'$ there exists an arc $\gamma \in \Gamma$ such that $h(\gamma) \subset \gamma'$. Thus $\mathrm{adm}(h(\Gamma)) \subset \mathrm{adm}(\Gamma')$, whence

$$\mathrm{mod}(h(\Gamma)) \geq \mathrm{mod}(\Gamma')$$

and

$$K^2 \geq \frac{\mathrm{mod}(h(\Gamma))}{\mathrm{mod}(\Gamma)} \geq \frac{2\,\pi - \alpha}{\alpha}\,\frac{\log(b/a)}{\log(b/a) + 2\log(c)}.$$

We conclude that

$$K^2 \geq \frac{2\,\pi - \alpha}{\alpha}$$

by letting $b/a \to \infty$.

Finally reversing the roles of $\mathbf{S}(\alpha)$ and $\mathbf{S}^*(\alpha)$ in the above argument yields

$$K^2 \geq \frac{\alpha}{2\,\pi - \alpha}$$

and hence (1.4.3).

DEFINITION 1.4.4. A domain D is a *sector of angle* α if it is the image of $\mathbf{S}(\alpha)$ under a similarity mapping.

Our second example is a simple Jordan domain that is not a quasidisk.

EXAMPLE 1.4.5. The half-strip

$$D = \{z = x + i\,y : 0 < x < \infty,\ |y| < 1\}$$

is not a quasidisk.

We shall show that there exists no quasiconformal self-mapping f of $\overline{\mathbf{R}}^2$ which maps $\mathbf{H}$ onto D. By performing a preliminary Möbius transformation, we need only consider the case where $f(\infty) = \infty$.

Suppose that f is a K-quasiconformal self-mapping of $\mathbf{R}^2$ with $f(\mathbf{H}) = D$, set $w_1 = x + i$, $w_2 = 0$, $w_3 = x - i$, and let $z_i = f^{-1}(w_i)$ for $i = 1, 2, 3$. Then z_1, z_2, z_3 is an ordered triple of points on $\partial\mathbf{H}$ with

$$|z_1 - z_2| < |z_1 - z_3|$$

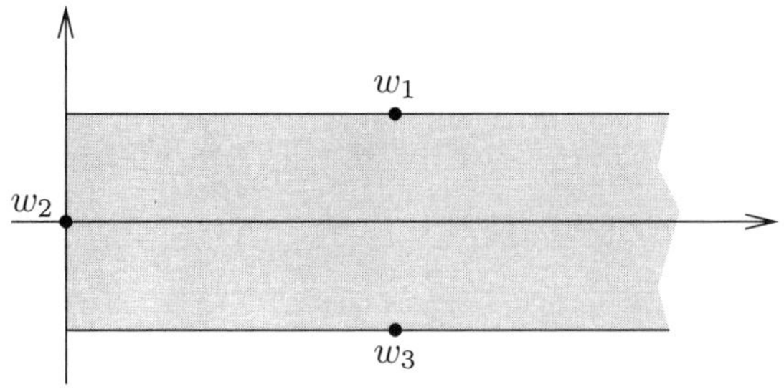

FIGURE 1.7

for each choice of x in $(0, \infty)$. On the other hand,

$$x < |f(z_1) - f(z_2)| \le c\,|f(z_1) - f(z_3)| = 2\,c$$

by Theorem 1.3.4 where $c = c(K)$ and we have a contradiction.

If D is a K-quasidisk, then ∂D is the image of a circle under a self-homeomorphism f of $\overline{\mathbf{R}}^2$ which is differentiable a.e. Thus ∂D is a Jordan curve which is a circle or line when $K = 1$. Hence it is natural to ask if ∂D has any nice analytic properties when $1 < K < \infty$. For example, is ∂D locally rectifiable?

Our third example shows that the answer is no and that, from the standpoint of Euclidean geometry, the boundary of a quasidisk can be quite wild. See Gehring-Väisälä [**70**].

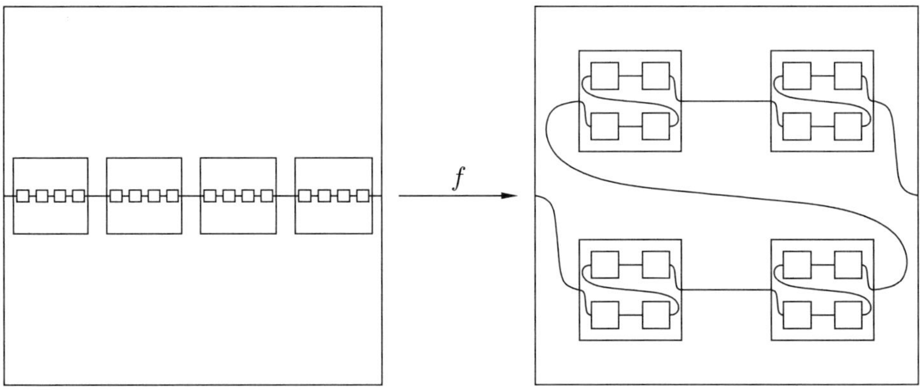

FIGURE 1.8

EXAMPLE 1.4.6. For each $1 < a < 2$ there exists a quasidisk D such that

$$\dim(\partial D) \ge a$$

where dim denotes Hausdorff dimension.

We will sketch a proof of this. We say that a square is *oriented* if its sides are parallel to the coordinate axes and we let Q and Q' denote the open squares

$$Q = Q' = \{z = x + i\,y : |x| < 1, |y| < 1\}.$$

Next set

$$z_1 = \frac{3}{4}, \qquad z_2 = \frac{1}{4}, \qquad z_3 = -\frac{1}{4}, \qquad z_4 = -\frac{3}{4}$$

and

$$w_1 = \frac{1+i}{2}, \qquad w_2 = \frac{-1+i}{2}, \qquad w_3 = \frac{1-i}{2}, \qquad w_4 = \frac{-1-i}{2},$$

and fix $0 < r < 1/2$. Then choose $0 < s < 1$ so that

$$\frac{\log 4}{\log(2/s)} = a.$$

Finally for $j = 1, 2, 3, 4$ let Q_j denote the open oriented square with center z_j and side length r and let Q'_j be the open oriented square with center w_j and side length s. Then we can choose a piecewise linear homeomorphism

$$f_0 : \overline{Q} \setminus \bigcup_{j=1}^{4} Q_j \to \overline{Q'} \setminus \bigcup_{j=1}^{4} Q'_j$$

such that f_0 is the identity on ∂Q and is of the form $a_j z + b_j, a_j > 0$, on ∂Q_j with $f_0(\partial Q_j) = \partial Q'_j$. Then f_0 is K-quasiconformal in $Q \setminus \bigcup_j \overline{Q}_j$ where $K = K(r, s)$.

Next for each j choose oriented squares $Q_{j,k}$ in Q_j and $Q'_{j,k}$ in Q'_j in the same way as the squares Q_j and Q'_j were chosen in Q and Q', respectively. By scaling we can extend f_0 to obtain a piecewise linear homeomorphism

$$f_1 : \overline{Q} \setminus \bigcup_{j,k=1}^{4} Q_{j,k} \to \overline{Q'} \setminus \bigcup_{j,k=1}^{4} Q'_{j,k}$$

which is K-quasiconformal in $Q \setminus \bigcup_{j,k} \overline{Q}_{j,k}$.

Continuing in this way, we obtain a homeomorphism

$$f : \overline{Q} \setminus E \to \overline{Q'} \setminus E'$$

where E and E' are Cantor sets. Then f can be extended by continuity to give a K-quasiconformal mapping which maps $\overline{Q}$ onto $\overline{Q'}$ and is the identity on ∂Q.

Set $f(z) = z$ in $\overline{\mathbf{R}}^2 \setminus \overline{Q}$. Then f is a K-quasiconformal self-mapping of $\overline{\mathbf{R}}^2$ which maps the upper half-plane $\mathbf{H}$ onto a quasidisk D with Hausdorff dimension

$$\dim(\partial D) \geq \frac{\log 4}{\log(2/s)} = a.$$

See Beardon [**17**] or page 67 in Mattila [**127**].

Although the Hausdorff dimension of the boundary ∂D of a quasidisk D can be arbitrarily close to 2, it always satisfies $m(\partial D) = 0$ where m is planar Lebesgue measure. This follows from Lusin's property (N) of quasiconformal mappings in Theorem 1.1.8. On the other hand, a result due to Astala [**13**] gives the estimate

$$\dim(\partial D) \leq \frac{2K}{K+1}$$

for the Hausdorff dimension of the boundary ∂D of a K-quasidisk D.

Our final example, or rather class of examples, in this section illustrates how quasidisks arise naturally in complex dynamics.

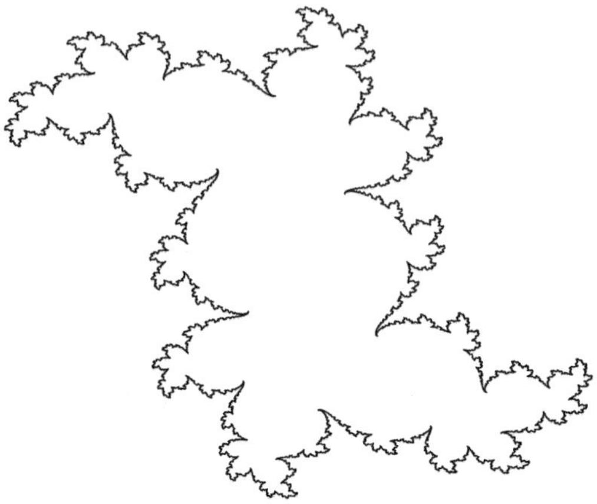

FIGURE 1.9

EXAMPLE 1.4.7. For a nonconstant meromorphic function $f\colon \overline{\mathbf{R}}^2 \to \overline{\mathbf{R}}^2$, the *iterates*

$$f^n(z) = f \circ f^{n-1}(z),\ n \geq 2, \qquad f^1(z) = f(z)$$

are all defined and meromorphic. The *Fatou set* F_f of f is the largest open set where the sequence (f^n) is a normal family, while its complement, $J_f = \overline{\mathbf{R}}^2 \setminus F_f$, is called the *Julia set*.

If p is a polynomial function of degree two, we may assume without loss of generality that it has the form

$$p_c(z) = z^2 + c.$$

If $c = 0$, the Julia set is the unit circle, and if $|c| < 1/4$, it can be shown that the Fatou set has exactly two components F_0 and F_∞, with $0 \in F_0$ and $\infty \in F_\infty$. See Beardon [**19**] or Carleson-Gamelin [**31**]. Arguments using Theorem 1.1.11 in an ingenious way reveal that in fact F_0 is a quasidisk. See e.g. Carleson-Gamelin [**31**].

1.5. What is ahead

Though quasidisks can be quite pathological domains, they occur very naturally in surprisingly many branches of analysis and geometry. We will describe in what follows some thirty different properties of quasidisks which generalize corresponding properties of Euclidean disks and which characterize this class of domains. See also Gehring [**51**] and [**54**].

The properties of a quasidisk D that we will discuss fall into the following categories:

1° geometric properties of D,
2° conformal invariants defined in D,
3° injectivity criteria for functions defined in D,
4° criteria for extension of functions defined in D,
5° two-sided criteria for D and D^*,
6° miscellaneous properties.

In the remainder of Part 1 (Chapters 2 to 7) we will consider properties in each of these categories. A number of them can be used to characterize Euclidean disks or half-planes. We will indicate when this is the case.

In Part 2 (Chapters 8 to 11) we will present proofs for some of the characterizations mentioned above. Many of the arguments follow a series of implications. There are four main series of implications, as well as some additional equivalences proved. Some proofs not in the main series of implications belong naturally to the discussion of the results and are presented in Part 1.

CHAPTER 2

Geometric properties

We begin our list of characterizations for a quasidisk D with six geometric properties. These include reflections in ∂D, reversed triangle inequalities for points in ∂D, a metric form of uniform local connectivity, and a decomposition property.

2.1. Reflection

Ahlfors introduced in [**5**] the concept of a quasiconformal reflection in a Jordan curve through ∞. We extend his definition to homeomorphic reflections in the boundary of an arbitrary domain.

DEFINITION 2.1.1. A domain $D \subset \overline{\mathbf{R}}^2$ admits a *reflection in its boundary* ∂D if there exists a homeomorphism of f of $\overline{D}$ such that

1° $f(D) = D^*$,
2° $f(z) = z$ for $z \in \partial D$.

This defines, in a natural way, a homeomorphism of $\overline{\mathbf{R}}^2$ which we will also denote by f.

The following result characterizes the plane domains which admit a reflection.

THEOREM 2.1.2. *A domain $D \subset \overline{\mathbf{R}}^2$ admits a reflection in ∂D if and only if it is a Jordan domain.*

PROOF. By performing a preliminary self-homeomorphism of $\overline{\mathbf{R}}^2$ we may assume without loss of generality that D is bounded.

If D is a Jordan domain, then by the Schönflies extension theorem there exists a self-homeomorphism g of $\overline{\mathbf{R}}^2$ which maps the unit disk $\mathbf{B}$ onto D. If r denotes reflection in $\partial\mathbf{B}$, then

$$f(z) = g \circ r \circ g^{-1}(z)$$

defines a self-homeomorphism of $\overline{\mathbf{R}}^2$ which satisfies 1° and 2°.

For the converse suppose that D admits a reflection f in ∂D. Then

$$D^* = f(D) \qquad \text{and} \qquad \partial D^* = f(\partial D) = \partial D. \tag{2.1.3}$$

We prove next that ∂D is locally connected at each of its points. Then since $\partial D = \partial D^*$, ∂D is a Jordan curve by Theorem IV.6.6 in Wilder [**166**].

Fix $z_0 \in \partial D$ and $\epsilon > 0$. Choose $\delta > 0$ so that $f(\mathbf{B}(z_0, \delta)) \subset \mathbf{B}(z_0, \epsilon)$ and suppose that

$$z_1, z_2 \in \partial D \cap \mathbf{B}(z_0, \delta).$$

We shall show that z_1 and z_2 lie in a connected set in $\partial D \cap \mathbf{B}(z_0, \epsilon)$ and hence that ∂D is locally connected at z_0.

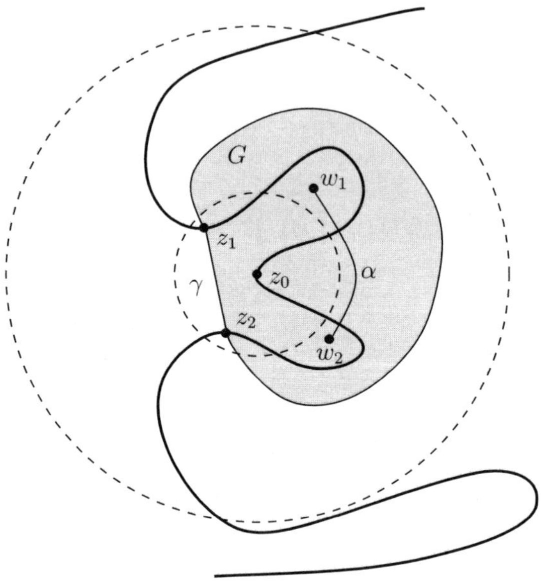

FIGURE 2.1

Let γ be the open segment (z_1, z_2) and suppose that $\gamma \cap \partial D = \emptyset$. Then

$$\gamma \cup f(\gamma) \cup \{z_1, z_2\}$$

is a Jordan curve which bounds a domain $G \subset \mathbf{B}(z_0, \epsilon)$. By our assumptions about f, $f(\overline{G}) = \overline{G}$.

If w_1, w_2 are points in $\overline{G} \cap \overline{D}$, then w_1 and w_2 can be joined by an arc α in $\overline{G}$ and hence by the continuum

$$\beta = (\alpha \cap \overline{D}) \cup f(\alpha \cap \overline{D^*}) \subset \overline{G} \cap \overline{D}.$$

Thus $\overline{G} \cap \overline{D}$ is connected. The same is true of $\overline{G} \cap \overline{D^*}$ and we conclude that

$$F_1 = \overline{G} \cap \overline{D}, \quad F_2 = \overline{G} \cap \overline{D^*}$$

are closed, connected sets with

$$F_1 \cap F_2 = \overline{G} \cap \partial D, \qquad F_1 \cup F_2 = \overline{G}.$$

Since $\overline{\mathbf{R}}^2 \setminus (F_1 \cup F_2)$ is a Jordan domain and hence connected, $E = F_1 \cap F_2$ is a connected set which joins z_1 and z_2 in $\partial D \cap \mathbf{B}(z_0, \epsilon)$. See, for example, Theorem V.11.5 in Newman [**140**].

Next if $\gamma \cap \partial D \neq \emptyset$, then

$$\gamma = (\cup_j \gamma_j) \cup (\gamma \cap \partial D)$$

where γ_j is an open segment (z_{1j}, z_{2j}) with $z_{1j}, z_{2j0} \in \partial D$ and $\gamma_j \cap \partial D = \emptyset$. Then as above there exists a connected set E_j which joins z_{1j} and z_{2j} in $\partial D \cap \mathbf{B}(z_0, \epsilon)$ and

$$E = (\cup_j E_j) \cup (\gamma \cap \partial D)$$

is a connected set which joins z_1 and z_2 in $\partial D \cap \mathbf{B}(z_0, \epsilon)$. Thus ∂D is locally connected at z_0. Since z_0 was chosen arbitrarily, ∂D itself is locally connected. $\square$

A domain $D \subset \overline{\mathbf{R}}^2$ is a Jordan domain if and only if it admits a reflection f in its boundary. What else can we say about D if we know more about the reflection f? When, for example, can we conclude that D is a quasidisk?

One obvious situation is as follows.

THEOREM 2.1.4 (Kühnau [**108**]). *A domain $D \subset \overline{\mathbf{R}}^2$ is a K-quasidisk if and only if it admits a K-quasiconformal reflection in ∂D.*

PROOF. If D admits a K-quasiconformal reflection f in ∂D, then D is a Jordan domain by Theorem 2.1.2. Hence there exists a homeomorphism h which maps $\overline{D}$ onto the upper half-plane $\overline{\mathbf{H}}$ and is conformal in D. In this case

$$g(z) = \begin{cases} h(z) & \text{if } z \in \overline{D}, \\ r \circ h \circ f^{-1}(z) & \text{if } z \in D^*, \end{cases}$$

where $r(z) = \overline{z}$, defines a K-quasiconformal self-mapping of $\overline{\mathbf{R}}^2$ with $g(D) = \mathbf{H}$ and D is a K-quasidisk.

If D is a K-quasidisk, then there exists a K-quasiconformal self-mapping g of $\overline{\mathbf{R}}^2$ which maps D onto the upper half-plane $\mathbf{H}$ and

$$f(z) = g^{-1} \circ r \circ g(z)$$

defines a K^2-quasiconformal reflection in ∂D.

A more detailed argument based on Theorem 1.1.11 shows that the mapping f can be replaced by a K-quasiconformal mapping f^* and hence that D is actually a K-quasidisk. See Kühnau [**108**]. □

COROLLARY 2.1.5. *A quasiconformal mapping between two quasidisks can be extended to a quasiconformal self-mapping of $\overline{\mathbf{R}}^2$.*

Another far less obvious situation concerns the case when the reflection f is bilipschitz with respect to the Euclidean metric.

DEFINITION 2.1.6. A mapping $f : E \to E'$ is an *L-bilipschitz mapping*, or more precisely a *Euclidean L-bilipschitz mapping*, of $E \subset \overline{\mathbf{R}}^2$ if

$$\frac{1}{L}\,|z_1 - z_2| \le |f(z_1) - f(z_2)| \le L\,|z_1 - z_2|$$

for $z_1, z_2 \in E \setminus \{\infty\}$ and $f(\infty) = \infty$ if $\infty \in E$.

If D is a half-plane, then there exists a 1-bilipschitz mapping of $\overline{\mathbf{R}}^2$ that maps D onto its exterior D^* and is the identity on ∂D. More generally a sector D of angle α admits an L-bilipschitz reflection where L is bounded by constants depending only on α [**59**]. The optimal bilipschitz reflection in a sector has been determined by J. Miller [**130**].

THEOREM 2.1.7 (Miller [**130**]). *If $D = \mathbf{S}(\alpha)$, $0 \le \alpha \le \pi$, then ∂D admits an optimal L-bilipschitz reflection f with $L = \cot\theta$, where θ is the unique angle such that $0 \le \theta < \alpha/2$, $\theta \le \pi/4$, and $\phi(\theta) = 1$, where*

$$\phi(t) = \frac{\pi + 2t - \alpha}{\alpha - 2t}\tan^2 t.$$

Moreover this reflection is unique.

If $\pi \le \alpha < 2\pi$, then the unique optimal reflection is given by f, where f^{-1} is the above optimal reflection for $\mathbf{S}(2\pi - \alpha)$.

Example 1.4.5 shows that the boundary ∂D of a quasidisk D can be an extremely complicated Jordan curve, one which *almost* has positive two-dimensional measure. It is therefore quite surprising that all such curves admit a bilipschitz reflection if $\infty \in \partial D$.

THEOREM 2.1.8 (Ahlfors [**5**]). *If $\infty \in \partial D$, then D is a K-quasidisk if and only if it admits a Euclidean L-bilipschitz reflection, where K and L depend only on each other.*

The conformal analogue of this result also characterizes the domains D which are half-planes. In particular, D is a half-plane if and only if it admits a bilipschitz reflection with $L = 1$ (Hag [**77**]). To see this, let f be a 1-bilipschitz reflection in ∂D. Fix $z_1 \in D$ and let $\zeta \in \partial D$, $z_1, \zeta \neq \infty$. Then

$$|f(z_1) - \zeta| = |f(z_1) - f(\zeta)| = |z_1 - \zeta|.$$

Thus $\partial D \subset J \cup \{\infty\}$ where J is the perpendicular bisector of the segment $[z_1, f(z_1)]$. On the other hand, let $z \in J$ and consider the broken line $\gamma = [z_1, z] \cup [z, f(z_1)]$. Then γ must meet ∂D, while by what we just observed $\gamma \cap \partial D \subset \gamma \cap J = \{\zeta\}$. This implies that $z \in \partial D$, hence $\partial D = J$ and D is a half-plane.

Finally a third situation concerns the case where the reflection f is bilipschitz with respect to the hyperbolic metric. If $D \subset \overline{\mathbf{R}}^2$ is a simply connected domain, then there exists a conformal mapping $f : D \to \mathbf{B}^2$ such that $f(z_1) = 0$ and $f(z_2) = r$ where $0 < r < 1$. The hyperbolic distance between z_1 and z_2 is then given by

$$h_D(z_1, z_2) = \log \frac{1+r}{1-r}.$$

We will discuss this metric in more detail in Chapter 3. We have given the above definition now so that we can introduce the notion of a hyperbolic bilipschitz reflection.

DEFINITION 2.1.9. If D and D' are simply connected domains in $\overline{\mathbf{R}}^2$, then $f : D \to D'$ is a *hyperbolic L-bilipschitz mapping* if

$$\frac{1}{L}\, h_D(z_1, z_2) \leq h_{D'}(f(z_1), f(z_2)) \leq L\, h_D(z_1, z_2) \tag{2.1.10}$$

for $z_1, z_2 \in D$.

THEOREM 2.1.11 (Ahlfors [**7**]). *A domain D is a K-quasidisk if and only if it admits a hyperbolic L-bilipschitz reflection, where K and L depend only on each other.*

The conformal analogue of this result characterizes the domains D which are disks or half-planes: D is a disk or half-plane if and only if it admits a hyperbolic L-bilipschitz reflection with $L = 1$.

2.2. The three-point condition

If D is a disk or half-plane, then for each pair of points $z_1, z_2 \in \partial D \setminus \{\infty\}$,

$$\min_{j=1,2} \operatorname{diam}(\gamma_j) \leq |z_1 - z_2|$$

where diam denotes Euclidean diameter and γ_1, γ_2 are the components of $\partial D \setminus \{z_1, z_2\}$. More generally if D is a sector of angle α where $0 < \alpha < \pi$, then

$$\min_{j=1,2} \operatorname{diam}(\gamma_j) \leq a\, |z_1 - z_2| \tag{2.2.1}$$

where

$$a = \begin{cases} \csc(\alpha) & \text{if } 0 < \alpha \le \pi/2, \\ 1 & \text{if } \pi/2 < \alpha < \pi. \end{cases}$$

The following counterpart of inequality (2.2.1) yields another characterization for the class of quasidisks.

DEFINITION 2.2.2. A Jordan domain D satisfies the *three-point condition* if there exists a constant $a \ge 1$ such that for each pair of points $z_1, z_2 \in \partial D \setminus \{\infty\}$,

$$\min_{j=1,2} \operatorname{diam}(\gamma_j) \le a\,|z_1 - z_2| \tag{2.2.3}$$

where γ_1, γ_2 are the components of $\partial D \setminus \{z_1, z_2\}$.

This condition derives its name from the fact that it implies that

$$|z_1 - z| \le a\,|z_1 - z_2| \tag{2.2.4}$$

for each point z in the component of $\partial D \setminus \{z_1, z_2\}$ with minimum diameter. Conversely (2.2.4) implies (2.2.3) with $2a$.

Ahlfors' well-known three-point condition of a quasidisk is as follows.

THEOREM 2.2.5 (Ahlfors [**5**]). *A Jordan domain D is a K-quasidisk if and only if it satisfies the three-point condition with constant a, where K and a depend only on each other.*

We see from the three-point condition that a quasidisk cannot have any inward or outward cusps.

REMARK 2.2.6. The fact that all quasidisks satisfy the three-point condition follows directly from Theorem 1.3.4.

To see this, let D be a K-quasidisk. Then there is a K-quasiconformal mapping of $\overline{\mathbf{R}}^2$ onto itself with $D = f(G)$, where G is a disk or a half-plane. If $f(\infty) \neq \infty$, there is a Möbius transformation ϕ with $\phi(\infty) = f^{-1}(\infty)$. Then $F = f \circ \phi$ is a K-quasiconformal mapping with $F(\phi^{-1}(G)) = D$ and $F(\infty) = \infty$, and D is the image of a disk or half-plane under F.

Next choose $z_1, z_2 \neq \infty$ in ∂D and let $\zeta_j = F^{-1}(z_j) \in \partial G$. Let γ' be the subarc of $\partial G \setminus \{\zeta_1, \zeta_2\}$ of smallest Euclidean diameter and let $\gamma = F(\gamma')$ be its image in ∂D. Then for each $\zeta \in \gamma'$ we must have that

$$|\zeta - \zeta_1| \le |\zeta_1 - \zeta_2|$$

because G is a disk or a half-plane. Then Theorem 1.3.4 yields

$$|F(\zeta) - F(\zeta_1)| \le e^{8K} |F(\zeta_1) - F(\zeta_2)|,$$

or

$$|z - z_1| \le e^{8K} |z_1 - z_2|,$$

for each $z \in \gamma$, which means that $\operatorname{diam}(\gamma) \le 2e^{8K}|z_1 - z_2|$. Thus D satisfies the three-point condition.

A different route to this implication will be presented in Chapter 8, where we will also prove the sufficiency part of Theorem 2.2.5.

The criterion in Theorem 2.2.5 is particularly useful when proving that a given domain is a quasidisk since the following result shows it is only necessary to verify inequality (2.2.3) for relatively few pairs of points $z_1, z_2 \in \partial D$.

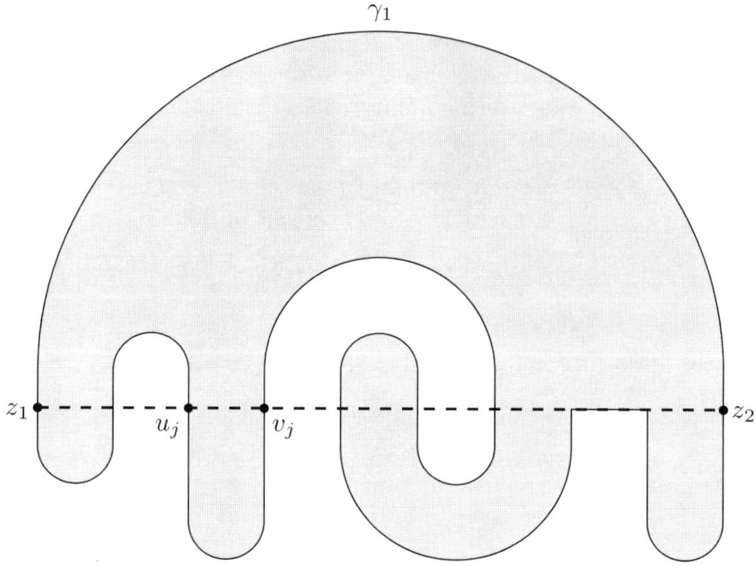

FIGURE 2.2

THEOREM 2.2.7 (Alestalo-Herron-Luukkainen [**10**]). *If inequality* (2.2.3) *holds for* $z_1, z_2 \in \partial D$ *such that*

$$\partial D \cap [z_1, z_2] = \{z_1, z_2\}, \tag{2.2.8}$$

then it holds for all $z_1, z_2 \in \partial D$ *with the constant* a *replaced by* $2a$.

PROOF. As before we let γ_1, γ_2 denote the components of $\partial D \setminus \{z_1, z_2\}$ and assume that $\text{diam}(\gamma_1) \leq \text{diam}(\gamma_2)$. More generally, for $u, v \in \partial D \cap [z_1, z_2]$ with $z_1 \leq u < v \leq z_2$ we let $\gamma_1(u, v)$ and $\gamma_2(u, v)$ denote the components of $\partial D \setminus \{u, v\}$ labeled so that $\text{diam}(\gamma_1(u, v)) \leq \text{diam}(\gamma_2(u, v))$.

Before we discuss the general case we consider the special case where $\overline{\gamma}_1 \cap [z_1, z_2] = \{z_1, z_2\}$. See Figure 2.2. Let $[u_j, v_j]$ denote the closures of the components of $[z_1, z_2] \setminus \overline{\gamma}_2$. Now $\partial D \cap [u_j, v_j] = \{u_j, v_j\}$ and (2.2.3) follows immediately if $\gamma_1(u_j, v_j) \supset \gamma_1$ for some j. Hence we assume $\gamma_1 \subset \gamma_2(u_j, v_j)$ for all j. In this case

$$\gamma_2 \subset [z_1, z_2] \cup \bigcup_j \gamma_1(u_j, v_j). \tag{2.2.9}$$

To see this, fix $z \in \gamma_2 \setminus [z_1, z_2]$ and let w_1 and w_2 be the first points of $\gamma_2 \cap [z_1, z_2]$ as γ_2 is traversed from z towards z_1 and z_2, respectively. We next consider the subarcs α and β say, of γ_2 joining z_1, w_1 and w_2, z_2, respectively. Then

$$\overline{\gamma}_2 \cap [z_1, z_2] = (\overline{\alpha} \cup \overline{\beta}) \cap [z_1, z_2].$$

Next select $u \in \overline{\alpha} \cap [z_1, z_2]$ and $v \in \overline{\beta} \cap [z_1, z_2]$ so that

$$|u - v| = \text{dist}(\overline{\alpha} \cap [z_1, z_2], \overline{\beta} \cap [z_1, z_2]).$$

Then the segment joining u and v (or v and u) equals $[u_j, v_j]$ for a suitable j. Since $\gamma_1 \subset \gamma_2(u_j, v_j)$, the ordering of the points implies that $z \in \gamma_1(u_j, v_j)$.

With (2.2.9) established, it is quite easy to finish the proof in the special case. In fact $\operatorname{diam}(\gamma_2) \leq a|z_1 - z_2|$ in this case. To see this, let $z, w \in \gamma_2$. If both w and z belong to $[z_1, z_2]$ or to some $\gamma_1(u_j, v_j)$, there is nothing to prove. If $w \in [z_1, z_2]$ and $z \notin [z_1, z_2]$, then $z \in \gamma_1(u_j, v_j)$ for some j. Now either $z_1 \leq w \leq u_j$ or $v_j \leq w \leq z_2$. We will assume the former, but a similar argument works for the second situation. Then

$$\begin{aligned}|w - z| &\leq |w - u_j| + |u_j - z| \\ &\leq |w - u_j| + a|u_j - v_j| \\ &\leq a(|w - u_j| + |u_j - v_j|) \\ &\leq a|z_1 - z_2|.\end{aligned}$$

Finally, suppose that $w \in \gamma_1(u_j, v_j)$ and $z \in \gamma_1(u_k, v_k)$, assuming $v_j \leq u_k$. Then

$$\begin{aligned}|w - z| &\leq |w - v_j| + |v_j - u_k| + |u_k - z| \\ &\leq a|u_j - v_j| + |v_j - u_k| + a|u_k - v_k| \\ &\leq a|u_j - v_k| \\ &\leq a|z_1 - z_2|.\end{aligned}$$

In the general case let $[z_{1k}, z_{2k}]$ denote the closures of the components of $[z_1, z_2] \setminus \overline{\gamma}_1$, and let $\gamma_{1k} = \gamma_1(z_{1k}, z_{2k}) \subset \gamma_1$. For each γ_{1k} we are in the situation we have treated already, and

$$|z_{1k} - z| \leq a\,|z_{1k} - z_{2k}|$$

for each $z \in \gamma_{1k}$. Finally $z \in \gamma_1$ must be in $[z_1, z_2]$ or in γ_{1k} for some k, in which case we have

$$\begin{aligned}|z_1 - z| &\leq |z_1 - z_{1k}| + |z_{1k} - z| \\ &\leq |z_1 - z_{1k}| + a\,|z_{1k} - z_{2k}| \\ &\leq a\,|z_1 - z_2|.\end{aligned}$$

By applying the triangle inequality we obtain (2.2.3) with the constant a replaced by $2a$. □

2.3. Reversed triangle inequality

Inequality (2.2.3) is not quantitatively sharp since for $a = 1$ it does not imply that D is a disk or half-plane. An alternative formulation of this condition as a reversed triangle inequality, or sum of two cross ratios, is better in this respect. The *cross ratio* of four points $z_1, z_2, z_3, z_4 \in \mathbf{R}^2$ is defined by

$$\frac{(z_1 - z_3)\,(z_2 - z_4)}{(z_1 - z_4)\,(z_2 - z_3)}.$$

When $z_4 = \infty$, the cross ratio reduces to

$$\frac{z_1 - z_3}{z_2 - z_3},$$

and similarly when z_1, z_2, or z_3 lies at ∞. See Ahlfors [**3**]. We will sometimes write $[z_1, z_2, z_3, z_4]$ for the cross ratio of $z_1, z_2, z_3, z_4 \in \overline{\mathbf{R}}^2$.

If D is a Jordan domain with $\infty \in \partial D$, then D satisfies the three-point condition if for some constant $b \geq 1$ each ordered triple of points $z_1, z_2, z_3 \in \partial D \setminus \{\infty\}$ satisfies the reversed triangle inequality

$$|z_1 - z_2| + |z_2 - z_3| \leq b\,|z_1 - z_3|,$$

whence

$$\frac{|z_1 - z_2|}{|z_1 - z_3|} + \frac{|z_2 - z_3|}{|z_1 - z_3|} \leq b.$$

When $\infty \notin \partial D$, the ratios on the left-hand side of the above inequality must be replaced by general cross ratios and we are led to the following alternative formulation of the three-point condition.

DEFINITION 2.3.1. A Jordan domain D satisfies the *reversed triangle inequality* if there exists a constant $b \geq 1$ such that

$$|z_1 - z_2||z_3 - z_4| + |z_2 - z_3||z_4 - z_1| \leq b\,|z_1 - z_3||z_2 - z_4|$$

for each ordered quadruple of points $z_1, z_2, z_3, z_4 \in \partial D \setminus \{\infty\}$.

LEMMA 2.3.2. *A Jordan domain D satisfies the reversed triangle inequality if and only if it satisfies the three-point condition.*

PROOF. Suppose that D satisfies the three-point condition with constant a and choose $z_1, z_2, z_3, z_4 \in \partial D \setminus \{\infty\}$. By relabeling if necessary, we may assume that

$$|z_1 - z_3| \leq |z_2 - z_4|.$$

Let γ_2 and γ_4 denote the components of $\partial D \setminus \{z_1, z_3\}$ which contain z_2 and z_4, respectively. Again by relabeling we may assume that

$$\operatorname{diam}(\gamma_2) \leq \operatorname{diam}(\gamma_4).$$

Then

$$|z_1 - z_2| \leq \operatorname{diam}(\gamma_2) \leq a\,|z_1 - z_3|, \qquad |z_2 - z_3| \leq \operatorname{diam}(\gamma_2) \leq a\,|z_1 - z_3|,$$

whence

$$|z_3 - z_4| \leq |z_2 - z_3| + |z_2 - z_4| \leq (a+1)|z_2 - z_4|,$$
$$|z_4 - z_1| \leq |z_1 - z_2| + |z_2 - z_4| \leq (a+1)|z_2 - z_4|.$$

Thus

$$|z_1 - z_2||z_3 - z_4| + |z_2 - z_3||z_4 - z_1| \leq b\,|z_1 - z_3||z_2 - z_4|$$

and D satisfies the reversed triangle inequality with constant $b = 2a(a+1)$.

Suppose next that D satisfies the reversed triangle inequality with constant b, fix $z_1, z_3 \in \partial D \setminus \{\infty\}$, and let γ_2 and γ_4 denote the components of $\partial D \setminus \{z_1, z_3\}$. If

$$\min_{j=2,4} \operatorname{diam}(\gamma_j) > 2b\,|z_1 - z_3|,$$

then we can choose $z_2 \in \gamma_2$ and $z_4 \in \gamma_4$ such that

$$|z_1 - z_2| > b\,|z_1 - z_3| \qquad \text{and} \qquad |z_1 - z_4| > b\,|z_1 - z_3|,$$

in which case

$$\begin{aligned} b\,|z_1 - z_3||z_2 - z_4| &\leq b\,|z_1 - z_3|(|z_2 - z_3| + |z_3 - z_4|) \\ &= b\,|z_1 - z_3||z_3 - z_4| + b\,|z_1 - z_3||z_2 - z_3| \\ &< |z_1 - z_2||z_3 - z_4| + |z_2 - z_3||z_4 - z_1|, \end{aligned}$$

a contradiction. Hence D satisfies the three-point condition with constant $2b$. $\square$

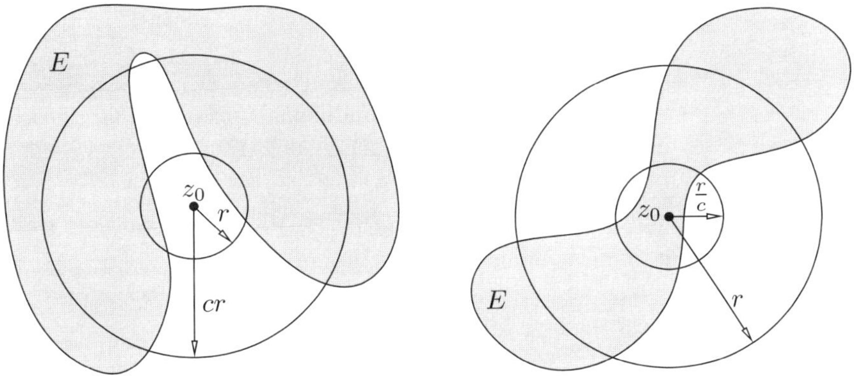

FIGURE 2.3

Thus a Jordan domain D is a quasidisk if and only if it satisfies the reversed triangle inequality. Moreover D is a disk or half-plane if and only if it satisfies the reversed triangle inequality with constant $b = 1$. See Ahlfors [**3**], Hag [**77**].

2.4. Linear local connectivity

We recall that a set $E \subset \overline{\mathbf{R}}^2$ is locally connected at a point $z_0 \in \overline{\mathbf{R}}^2$ if for each neighborhood U of z_0 there exists a second neighborhood V of z_0 such that $E \cap V$ lies in a component of $E \cap U$. Note that if $z_0 \in E$, this definition agrees with the intrinsic definition of local connectivity.

Now let D be a Jordan domain. The three-point condition (2.2.3) implies that the boundary $E = \partial D$ is locally connected where the sizes of U and V satisfy the following linear relations:

1° For each neighborhood $U = \mathbf{B}(z_0, r)$ of $z_0 \in \mathbf{R}^2$ we may choose $V = \mathbf{B}(z_0, s)$ where $s = r/(2a+1)$.

2° For each neighborhood $U = \overline{\mathbf{R}}^2 \setminus \overline{\mathbf{B}}(z_1, r)$ of $z_0 = \infty$ we may choose $V = \overline{\mathbf{R}}^2 \setminus \overline{\mathbf{B}}(z_1, s)$ where $s = (2a+1)r$.

Conversely, conditions 1° and 2° together imply that the three-point condition holds. This observation suggests the following property which also characterizes the class of quasidisks when E is a simply connected domain.

DEFINITION 2.4.1. A set $E \subset \overline{\mathbf{R}}^2$ is *linearly locally connected* if there exists a constant $c \geq 1$ such that

1° $E \cap \mathbf{B}(z_0, r)$ lies in a component of $E \cap \mathbf{B}(z_0, cr)$ and

2° $E \setminus \overline{\mathbf{B}}(z_0, r)$ lies in a component of $E \setminus \overline{\mathbf{B}}(z_0, r/c)$

for each $z_0 \in \mathbf{R}^2$ and each $r > 0$.

It follows from the definition that E itself is connected. By condition 1° $E \setminus \{\infty\}$ is connected since two points from different components cannot be joined in any disk. If E contains ∞, then ∞ must be an accumulation point by condition 2°, and E is connected since $E \setminus \{\infty\}$ is connected.

REMARK 2.4.2. Condition 2° in Definition 2.4.1 holds if condition 1° holds for E and its image under each Möbius transformation $f : \overline{\mathbf{R}}^2 \to \overline{\mathbf{R}}^2$.

PROOF. Choose $z_1, z_2 \in E \setminus \overline{\mathbf{B}}(z_0, r)$ and let

$$f(z) = r^2 \frac{z - z_0}{|z - z_0|^2} + z_0.$$

Then $f(z_1), f(z_2) \in f(E) \cap \overline{\mathbf{B}}(z_0, r)$. Next condition 1° implies that these points can be joined by a connected set γ in $f(E) \cap \overline{\mathbf{B}}(z_0, cr)$. Hence $f^{-1}(\gamma)$ joins z_1 and z_2 in $E \setminus \mathbf{B}(z_0, r/c)$ and condition 2° holds. □

The connection cited above between linear local connectivity and the three-point condition yields the following reformulation of Theorem 2.2.5.

COROLLARY 2.4.3 (Walker [**165**]). *A Jordan domain D is a K-quasidisk if and only if ∂D is linearly locally connected with constant c, where K and c depend only on each other.*

In contrast to the three-point condition, the property of linear local connectivity can be applied to a sdomain as well as to its boundary. For example, a sector D of angle α is linearly locally connected with $c = \csc(\alpha/2)$.

The converse of the Jordan curve theorem implies that a simply connected domain D is a Jordan domain if and only if D is locally connected at each point of $\overline{\mathbf{R}}^2$ (Newman [**140**]).

We have the following counterpart of this result for quasidisks.

THEOREM 2.4.4 (Gehring [**49**]). *A simply connected domain D is a K-quasidisk if and only if it is linearly locally connected with constant c, where K and c depend only on each other.*

Moreover D is a disk or half-plane if and only if it is linearly locally connected with constant $c = 1$. See Langmeyer [**110**].

Finally we show that in general, the notion of linear local connectivity is invariant with respect to Möbius transformations. Our proof depends on the following result.

LEMMA 2.4.5. *Suppose that $1 < c < \infty$ and that $E_1, E_2 \subset \overline{\mathbf{R}}^2$ are sets which are separated by an annulus*

$$\{z : a < |z - z_0| < ac\}.$$

If f is a Möbius transformation, then $f(E_1)$ and $f(E_2)$ are separated by an annulus

$$\{w : b < |w - w_0| < bg(c)\}$$

where $g(c) = c^{1/2} + c^{-1/2} - 1$. This bound is sharp.

PROOF. By means of a preliminary similarity mapping we may assume that $a = 1$ and $z_0 = 0$. Let

$$C_1 = \{z : |z| = 1\}, \qquad C_2 = \{z : |z| = c\}.$$

Since the inversion $i(z) = c/z$ interchanges C_1 and C_2, we may also assume that

$$\operatorname{diam} f(C_1) \le \operatorname{diam} f(C_2). \tag{2.4.6}$$

This implies that $f(C_1)$ is a circle and $f(C_2)$ is a circle or line. Hence there exists a line L' through the center of $f(C_1)$ which contains the center of $f(C_2)$ if $f(C_2)$ is a circle or is orthogonal to $f(C_2)$ if $f(C_2)$ is a line. Now $L = f^{-1}(L')$ is a line through the origin which is orthogonal to the concentric circles C_1 and C_2.

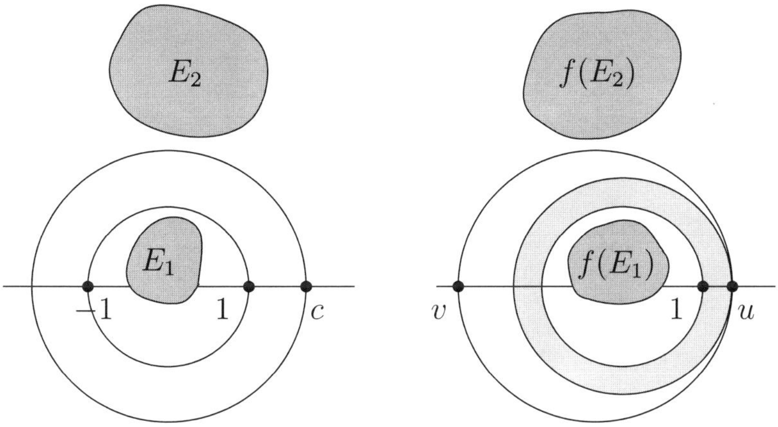

FIGURE 2.4

By performing a final pair of similarity mappings we may assume that L' as well as L is the real axis and that

$$f(1) = 1, \qquad f(-1) = -1, \qquad f(c) = u > 1, \qquad f(-c) = v$$

and it is enough to consider the cases where $f(-c) = v$ where $v \leq -u$ or $v \geq u + 2$ including $v = \infty$. Then $f(E_1)$ and $f(E_2)$ will be separated by the annulus $\{w : 1 < |w| < u\}$. Finally, using the fact that cross ratios are invariant under Möbius transformations we have that

$$(c^{1/2} + c^{-1/2})^2 = 4\frac{(c+1)(1+c)}{(c+c)(1+1)} = 4\frac{(u+1)(1-v)}{(u-v)(1+1)} = h(u,v)$$

where

$$h(u,v) = 2\frac{1-uv}{u-v} + 2.$$

Since

$$\frac{\partial h(u,v)}{\partial v} = 2\frac{1-u^2}{(u-v)^2} < 0,$$

$h(u,v)$ is decreasing in v, and we have that

$$(c^{1/2} + c^{-1/2})^2 \leq \max_{v \leq -u} h(u,v) = \lim_{v \to -\infty} h(u,v) = 2(u+1) < (u+1)^2$$

if $v \leq -u$ and

$$(c^{1/2} + c^{-1/2})^2 \leq \max_{v \geq u+2} h(u,v) = \lim_{v \to u+2} h(u,v) = (u+1)^2$$

if $v \geq u + 2$. In either case,

$$g(c) = c^{1/2} + c^{-1/2} - 1 \leq u$$

and hence $f(E_1)$ and $f(E_2)$ are separated by $\{w : 1 < |w| < g(c)\}$.

The extremal situation occurs when

$$f(1) = 1, \qquad f(-1) = -1, \qquad f(c) = g(c), \qquad f(-c) = g(c) + 2,$$

in which case

$$f(C_1) = \{z : |z| = 1\}, \qquad f(C_2) = \{z : |z - (g(c)+1)| = 1\}.$$

□

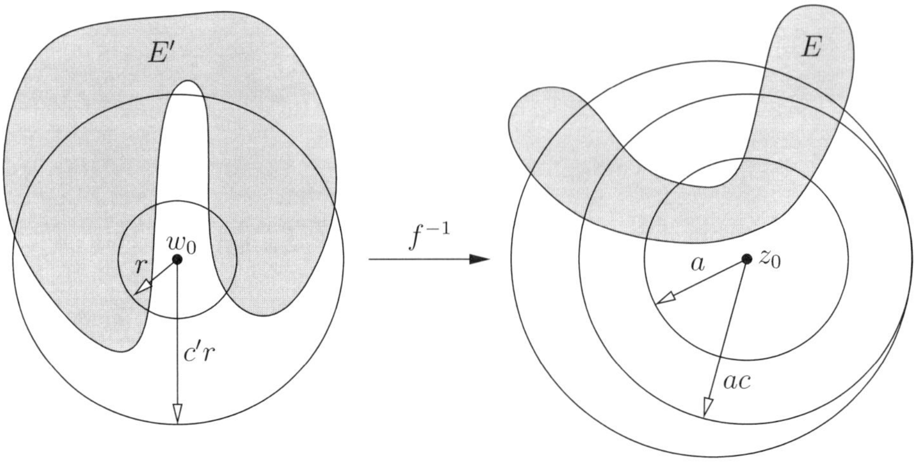

FIGURE 2.5

THEOREM 2.4.7 (Gehring [**48**], Walker [**165**]). *The image of a c-linearly locally connected set* $E \subset \overline{\mathbf{R}}^2$ *under a Möbius transformation* f *is* c'*-linearly locally connected where* $c' = g^{-1}(c)$ *and*

$$g(t) = t^{1/2} + t^{-1/2} - 1.$$

PROOF. By Remark 2.4.2 it is enough to show that $E' = f(E)$ satisfies condition 1°. So assume that this is not true, i.e. that there exist $w_0 \in \mathbf{R}^2$ and $r > 0$ so that $E'_1 = E' \cap \mathbf{B}(w_0, r)$ does not lie in a component of $E' \cap \mathbf{B}(w_0, c'r)$. Let C' be an arbitrary connected subset of E' containing E'_1 and set

$$E'_2 = C' \setminus \mathbf{B}(w_0, c'r) \neq \emptyset.$$

Then $C = f^{-1}(C')$ is an arbitrary connected set in E containing $E_1 = f^{-1}(E'_1)$ and $E_2 = f^{-1}(E'_2) \subset C$. By Lemma 2.4.5, E_1 and E_2 are separated by an annulus

$$\{z\colon a < |z - z_0| < ca\}$$

in contradiction to the hypothesis that E is c-linearly locally connected. □

2.5. Decomposition

The following observation is the basis for another property of quasidisks.

EXAMPLE 2.5.1. A domain $D \subsetneq \mathbf{R}^2$ is a disk or half-plane if and only if for each pair of points $z_1, z_2 \in D$ there exists a disk D' with

$$z_1, z_2 \in D' \subset D.$$

To see this, we need only prove the sufficiency, and we consider two cases. If D is bounded, we can choose z_{1j} and z_{2j} in D so that

$$|z_{1j} - z_{2j}| \to \operatorname{diam}(D) = 2\,r$$

as $j \to \infty$. By hypothesis, there exists for each j a disk $B_j = \mathbf{B}(w_j, r_j)$ with $z_{1j}, z_{2j} \in B_j \subset D$ and, by passing to a subsequence, we may assume that $w_j \to z_0 \in$

$\overline{D}$ and $r_j \to r$ as $j \to \infty$. Hence if $z \in B_0 = \mathbf{B}(z_0, r)$, then for large j

$$|z - w_j| \leq |z - z_0| + |z_0 - w_j| < r_j,$$

and $z \in B_j \subset D$. Thus $B_0 \subset D$, $\operatorname{diam}(B_0) = \operatorname{diam}(D)$, and $B_0 = D$.

If D is unbounded, then there exists a point $z_0 \in \mathbf{R}^2 \setminus D$. Next since D is convex, $2z_0 - z \notin D$ whenever $z \in D$. Thus $D \cap \phi(D) = \emptyset$ when $\phi(z) = 2z_0 - z$ and $\phi(D)$ is an open set. Let f be a Möbius transformation

$$f(z) = \frac{1}{z - \phi(z_1)}$$

with $z_1 \in D$. Then $f(D)$ is bounded and $f(D)$ is a disk by what was proved above. We conclude that D is a disk or half-plane.

DEFINITION 2.5.2. A domain D is *K-quasiconformally decomposable* if for each $z_1, z_2 \in D$ there exists a K-quasidisk D' with

$$z_1, z_2 \in D' \subset D.$$

The argument in Example 2.5.1 required only the existence of a point $z_0 \in \mathbf{R}^2 \setminus D$. However, the hypothesis that D is simply connected is essential in a proof of the corresponding result for quasidisks. For example, if $D = \mathbf{R}^2 \setminus \{0\}$, then D is K-quasiconformally decomposable for each $K > 1$ but D is clearly not a quasidisk.

THEOREM 2.5.3 (Gehring-Osgood [**67**]). *A simply connected domain D is a K-quasidisk if and only if it is K'-quasiconformally decomposable, where K and K' depend only on each other.*

CHAPTER 3

Conformal invariants

We consider here eight conformally invariant descriptions for a quasidisk D. Four of these compare the hyperbolic and Euclidean geometries in D, two relate the harmonic measures of arcs in ∂D at points in D and D^*, one compares the moduli of quadrilaterals in D and D^* with common vertices in ∂D, and the last relates the extremal distance between pairs of continua in D and in $\overline{\mathbf{R}}^2$.

3.1. Conformal invariants in a Jordan domain

A Jordan domain D together with a finite set of interior points $z_1, \ldots, z_m \in D$ and boundary points $w_1, \ldots, w_n \in \partial D$ taken in a certain order is said to be a *configuration* Σ. Two configurations are *conformally equivalent* if there exists a conformal mapping of one domain onto the other which maps specified interior and boundary points of one domain onto corresponding points of the other.

Suppose Σ' is a second configuration consisting of a Jordan domain D' with points $z'_1, \ldots, z'_m \in D'$ and $w'_1, \ldots, w'_n \in \partial D'$. Then since there exist homeomorphisms

$$f : \overline{D} \to \overline{\mathbf{B}} \qquad \text{and} \qquad g : \overline{D'} \to \overline{\mathbf{B}}$$

that are conformal in D and D', respectively, in order to decide if Σ is conformally equivalent to Σ' it is sufficient to consider the case where $D = D' = \mathbf{B}$. In this case Σ and Σ' are each determined by $2m + n$ real numbers.

Since $\mathbf{B}$ has conformal self-mappings which carry any pair of interior and boundary points z_1, w_1 onto any other pair z'_1, w'_1 or, alternatively, any ordered triple of boundary points w_1, w_2, w_3 onto any other ordered triple w'_1, w'_2, w'_3, the

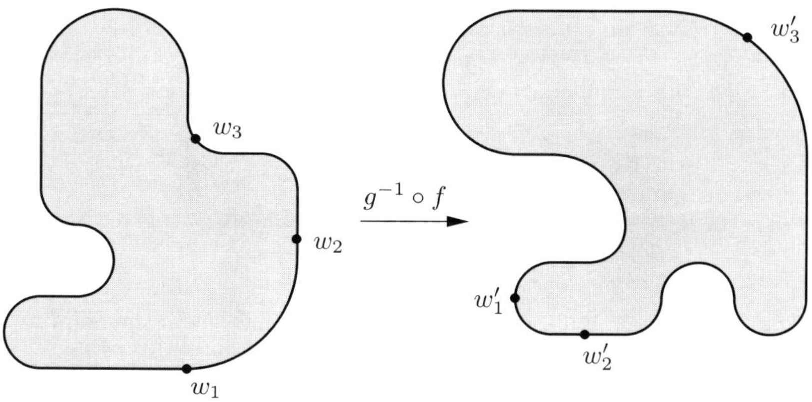

FIGURE 3.1

conformal type of Σ is determined by

$$N = 2m + n - 3$$

real numbers. The three cases where $N = 1$, together with a natural real number or *conformal invariant* that determines the conformal equivalence class of Σ, are as follows. See Ahlfors [**8**].

1° Two interior points z_1, z_2. The conformal invariant is the hyperbolic distance $h_D(z_1, z_2)$ between z_1 and z_2.
2° One interior point z_1 and two boundary points w_1, w_2. The conformal invariant is the harmonic measure $\omega(z_1, \gamma; D)$ where γ is one of the boundary arcs with endpoints w_1, w_2.
3° Four boundary points $w_1, \ldots, w_4$. The conformal invariant is the modulus of the quadrilateral $Q = D$ with vertices at $w_1, \ldots, w_4$.

We shall describe in this chapter how each of these invariants can be used to characterize the class of quasidisks.

3.2. Hyperbolic geometry

We begin by defining in each simply connected domain $D \subset \overline{\mathbf{R}}^2$ a conformally invariant distance, the *hyperbolic metric*, or *hyperbolic distance*, h_D.

The *hyperbolic density* in the unit disk $\mathbf{B}$ is defined by

$$\rho_{\mathbf{B}}(z) = \frac{2}{1 - |z|^2}$$

for $z \in \mathbf{B}$. Next for each $z_1, z_2 \in \mathbf{B}$ the *hyperbolic distance* between these points is given by

$$h_{\mathbf{B}}(z_1, z_2) = \inf_{\beta} \int_{\beta} \rho_{\mathbf{B}}(z)|dz|,$$

where the infimum is taken over all rectifiable curves β which join z_1 and z_2 in $\mathbf{B}$. Then there exists a unique arc γ such that

$$h_{\mathbf{B}}(z_1, z_2) = \int_{\gamma} \rho_{\mathbf{B}}(z)|dz|.$$

The arc γ lies in the circular crosscut β of $\mathbf{B}$ which passes through z_1, z_2 and is orthogonal to $\partial\mathbf{B}$. We call γ the *hyperbolic segment* joining z_1 and z_2 and we call β the *hyperbolic line* which contains γ. It is not difficult to show that

$$h_{\mathbf{B}}(z_1, z_2) = \log\left(\frac{|1 - \bar{z}_1 z_2| + |z_1 - z_2|}{|1 - \bar{z}_1 z_2| - |z_1 - z_2|}\right).$$

The *hyperbolic density* in a simply connected domain D is given by

$$\rho_D(z) = \rho_{\mathbf{B}}(g(z))|g'(z)|,$$

where g is any conformal mapping of D onto $\mathbf{B}$. It follows from the Schwarz lemma that ρ_D is independent of the choice of g. Then the *hyperbolic distance* between $z_1, z_2 \in D$ is given by

$$h_D(z_1, z_2) = \inf_{\beta} \int_{\beta} \rho_D(z)|dz|,$$

where the infimum is taken over all rectifiable curves β which join z_1 and z_2 in D. Again there is a unique *hyperbolic segment* γ in D for which

$$h_D(z_1, z_2) = \int_\gamma \rho_D(z)|dz|.$$

Then

$$h_D(z_1, z_2) = h_{\mathbf{B}}(g(z_1), g(z_2))$$

and g preserves the class of hyperbolic segments in D and $\mathbf{B}$.

Finally from the Schwarz lemma and the Koebe distortion theorem (Ahlfors [**8**]), it follows that

$$\frac{1}{2\,\mathrm{dist}(z, \partial D)} \le \rho_D(z) \le \frac{2}{\mathrm{dist}(z, \partial D)} \tag{3.2.1}$$

for $z \in D$, where $\mathrm{dist}(z, \partial D)$ denotes the Euclidean distance from z to ∂D. In particular,

$$\rho_D(z) = \frac{1}{\mathrm{Im}(z)} = \frac{1}{\mathrm{dist}(z, \partial D)}$$

if D is the upper half-plane $\mathbf{H}$ and

$$\rho_D(r\,e^{i\,\theta}) = \frac{\pi}{\alpha}\,\sec\left(\frac{\pi\,\theta}{\alpha}\right)\frac{1}{r} \tag{3.2.2}$$

if $D = \mathbf{S}(\alpha)$.

We show in what follows how a quasidisk D can be characterized in four different ways by comparing the Euclidean and hyperbolic geometries in D and its exterior D^*:

1° Bound for hyperbolic distance. This asserts that the hyperbolic distance between points in D is bounded above by a function of the ratio of the Euclidean distance between the points and their Euclidean distances from ∂D.

2° Geometry of hyperbolic segments. This describes the Euclidean length and position of hyperbolic segments in D in terms of their endpoints.

3° Min-max property of hyperbolic segments. This assumes that, up to a constant factor, the endpoints of hyperbolic segments γ in D minimize and maximize the Euclidean distance between points in γ and in D^*.

4° Hyperbolic reflection. This asserts the existence of a reflection in ∂D that is bilipschitz with respect to the hyperbolic distances in D and D^*.

3.3. Bounds for hyperbolic distance

The hyperbolic distance h_D is, by its nature, difficult to calculate in all but the simplest of domains. However the following two metrics yield lower bounds for h_D when D is simply connected. The first is the *Apollonian metric*

$$a_D(z_1, z_2) = \sup_{w_1, w_2 \in \partial D} \log\left(\frac{|z_1 - w_1||z_2 - w_2|}{|z_2 - w_1||z_1 - w_2|}\right), \tag{3.3.1}$$

see Beardon [**20**] and Gehring-Hag [**58**], and the second is the *distance-ratio metric*

$$j_D(z_1, z_2) = \log\left(\frac{|z_1 - z_2|}{\mathrm{dist}(z_1, \partial D)} + 1\right)\left(\frac{|z_1 - z_2|}{\mathrm{dist}(z_2, \partial D)} + 1\right), \tag{3.3.2}$$

see Gehring-Palka [**68**]. Moreover, these metrics also furnish upper bounds for h_D whenever D is a quasidisk.

LEMMA 3.3.3. *The Apollonian metric a_D is a Möbius invariant pseudo-metric in D. It is a metric whenever ∂D is not a proper subset of a circle or line.*

PROOF. Equation (3.3.1) implies that a_D is a pseudo-metric which is invariant with respect to Möbius transformations; a_D is a metric if, in addition, $a_D(z_1, z_2) > 0$ whenever z_1 and z_2 are distinct points of D. By the Möbius invariance we may assume that $z_1 = 0$ and $z_2 = \infty$. In this case

$$a_D(0, \infty) = \sup_{w_1, w_2 \in \partial D} \log\left(\frac{|w_1|}{|w_2|}\right) = 0$$

if and only if ∂D lies in a circle about 0. □

LEMMA 3.3.4. *The distance-ratio metric j_D is a similarity invariant metric in $D \subset \mathbf{R}^2$.*

PROOF. Equation (3.3.2) implies that j_D is invariant with respect to similarity transformations. To complete the proof, it suffices to show that

$$l = j_D(z_1, z_3) \leq j_D(z_1, z_2) + j_D(z_2, z_3) = r$$

for $z_1, z_2, z_3 \in D$. Let $d_i = \operatorname{dist}(z_i, \partial D)$ for $i = 1, 2, 3$. Then from the inequalities

$$\frac{|z_1 - z_2| + d_2}{d_2} \geq \frac{|z_1 - z_2| + |z_2 - z_3| + d_3}{|z_2 - z_3| + d_3} \geq \frac{|z_1 - z_3| + d_3}{|z_2 - z_3| + d_3}$$

and

$$\frac{|z_2 - z_3| + d_2}{d_2} \geq \frac{|z_2 - z_3| + |z_1 - z_2| + d_1}{|z_1 - z_2| + d_1} \geq \frac{|z_1 - z_3| + d_1}{|z_1 - z_2| + d_1}$$

we obtain

$$\begin{aligned}
\exp(r) &= \frac{|z_1 - z_2| + d_1}{d_1} \, \frac{|z_1 - z_2| + d_2}{d_2} \, \frac{|z_2 - z_3| + d_2}{d_2} \, \frac{|z_2 - z_3| + d_3}{d_3} \\
&\geq \frac{|z_1 - z_2| + d_1}{d_1} \, \frac{|z_1 - z_3| + d_3}{|z_2 - z_3| + d_3} \, \frac{|z_1 - z_3| + d_1}{|z_1 - z_2| + d_1} \, \frac{|z_2 - z_3| + d_3}{d_3} \\
&= \frac{|z_1 - z_3| + d_1}{d_1} \, \frac{|z_1 - z_3| + d_3}{d_3} = \exp(l).
\end{aligned}$$

□

The following two results indicate how a_D and j_D are related to the hyperbolic metric h_D in a simply connected domain.

LEMMA 3.3.5 (Gehring-Palka [**68**]). *If D is simply connected, then*

$$a_D(z_1, z_2) \leq j_D(z_1, z_2) \leq 4\, h_D(z_1, z_2) \tag{3.3.6}$$

for $z_1, z_2 \in D$.

PROOF. Fix z_1, z_2 in D and $w_1, w_2 \in \partial D$. Then

$$\frac{|z_1 - w_j|}{|z_2 - w_j|} \leq \frac{|z_1 - z_2| + |z_2 - w_j|}{|z_2 - w_j|} \leq \frac{|z_1 - z_2|}{\operatorname{dist}(z_2, \partial D)} + 1$$

for $j = 1, 2$, whence

$$\frac{|z_1 - w_1||z_2 - w_2|}{|z_2 - w_1||z_1 - w_2|} \leq \left(\frac{|z_1 - z_2|}{\operatorname{dist}(z_1, \partial D)} + 1\right)\left(\frac{|z_1 - z_2|}{\operatorname{dist}(z_2, \partial D)} + 1\right)$$

and the first part of (3.3.6) follows from (3.3.1) and (3.3.2).

Next let γ be the hyperbolic segment joining z_1 and z_2 in D. Then

$$\frac{1}{\operatorname{dist}(z_j, \partial D) + |z - z_j|} \leq \frac{1}{\operatorname{dist}(z, \partial D)} \leq 2\, \rho_D(z)$$

for $z \in \gamma$ by (3.2.1) and thus

$$\log\left(\frac{|z_1 - z_2|}{\operatorname{dist}(z_j, \partial D)} + 1\right) = \int_\gamma \frac{d\,|z - z_j|}{\operatorname{dist}(z_j, \partial D) + |z - z_j|} \leq 2\, h_D(z_1, z_2).$$

Adding these inequalities for $j = 1, 2$ yields the second part of (3.3.6). □

EXAMPLE 3.3.7. If D is a disk or half-plane, then

$$h_D(z_1, z_2) = a_D(z_1, z_2) \leq j_D(z_1, z_2) \tag{3.3.8}$$

for $z_1, z_2 \in D$. The second part of (3.3.8) holds with equality whenever D is a disk and z_1, z_2 lie on a diameter and are separated by the center of D.

Since a_D and h_D are invariant with respect to Möbius transformations, we may assume that $D = \mathbf{B}$ and that $z_1 = 0$ and $z_2 = r$ where $0 < r < 1$ in the proof for the first part of (3.3.8). Then

$$a_{\mathbf{B}}(0, r) = \sup_{|w_1| = |w_2| = 1} \log\left(\frac{|w_1 - r|}{|w_2 - r|}\right) = \log\left(\frac{1 + r}{1 - r}\right) = h_{\mathbf{B}}(0, r).$$

The inequality in the second part of (3.3.8) follows from Lemma 3.3.5.

Next since a_D and j_D are invariant with respect to similarity mappings, we may assume that $D = \mathbf{B}$ and that $z_1 = r_1 < 0$ and $z_2 = r_2 > 0$ for the proof of equality in the second part (3.3.8) in which case

$$\begin{aligned} a_{\mathbf{B}}(r_1, r_2) &= \log\left(\frac{|r_1 - 1|}{|r_1 + 1|}\,\frac{|r_2 + 1|}{|r_2 - 1|}\right) \\ &= \log\left(\frac{|r_2 - r_1|}{\operatorname{dist}(r_1, \partial D)} + 1\right)\left(\frac{|r_2 - r_1|}{\operatorname{dist}(r_2, \partial D)} + 1\right) = j_{\mathbf{B}}(r_1, r_2). \end{aligned}$$

The following counterpart of Example 3.3.7 for the case where D is an angular sector suggests what happens when D is a quasidisk.

EXAMPLE 3.3.9. If $D = \mathbf{S}(\alpha)$ where $0 < \alpha \leq \pi$, then

$$h_D(z_1, z_2) \leq \frac{\pi}{\alpha}\, a_D(z_1, z_2) \leq \frac{\pi}{\alpha}\, j_D(z_1, z_2) \tag{3.3.10}$$

for $z_1, z_2 \in D$ whenever $|z_1| = |z_2|$ or z_1/z_2 is real. The bound $\dfrac{\pi}{\alpha}$ is sharp.

For this suppose that $z_j = r\, e^{i\,\theta_j}$ where $-\alpha/2 < \theta_1 < \theta_2 < \alpha/2$. Then

$$a_D(z_1, z_2) \geq \log\left(\frac{|z_1 - w_2||z_2 - w_1|}{|z_1 - w_1||z_2 - w_2|}\right), \tag{3.3.11}$$

where $w_1 = r\,e^{-i\,\alpha/2}$ and $w_2 = r\,e^{i\,\alpha/2}$, and with (3.2.2) we obtain

$$\begin{aligned} a_D(z_1,z_2) &\geq \frac{1}{2}\int_{\theta_1}^{\theta_2}\left(\cot\frac{\alpha-2\,\theta}{4}+\cot\frac{\alpha+2\,\theta}{4}\right)d\theta \\ &= \int_{\theta_1}^{\theta_2}\frac{\sin(\alpha/2)}{\cos\theta-\cos(\alpha/2)}\,d\theta \\ &\geq \int_{\theta_1}^{\theta_2}\sec\left(\frac{\pi\,\theta}{\alpha}\right)d\theta = \frac{\alpha}{\pi}\int_{\theta_1}^{\theta_2}\rho_D(re^{i\,\theta})\,rd\theta \\ &\geq \frac{\alpha}{\pi}\,h_D(z_1,z_2). \end{aligned}$$

Suppose next that $z_j = r_j\,e^{i\,\theta}$ where $0 < r_1 < r_2$ and $-\alpha/2 < \theta < \alpha/2$. If $\theta = 0$, then (3.3.11) with $w_1 = 0$ and $w_2 = \infty$ yields

$$\begin{aligned} a_D(z_1,z_2) &\geq \log\left(\frac{|z_2|}{|z_1|}\right) = \int_{r_1}^{r_2}\frac{dr}{r} \\ &= \frac{\alpha}{\pi}\int_{r_1}^{r_2}\rho_D(r)\,dr \geq \frac{\alpha}{\pi}\,h_D(z_1,z_2). \end{aligned}$$

A more detailed calculation yields the same result when $\theta \neq 0$. These estimates are asymptotically sharp in each case as $h_D(z_1,z_2)\to\infty$.

From Example 3.3.9 it follows that

$$h_D(z_1,z_2) \leq \frac{2\pi}{\alpha}\,a_D(z_1,z_2) \leq \frac{2\pi}{\alpha}\,j_D(z_1,z_2)$$

for $z_1, z_2 \in D$ when D is a sector of angle α with $0 < \alpha \leq \pi$. The following results show that similar bounds hold whenever D is a quasidisk.

THEOREM 3.3.12 (Jones [**94**]). *A simply connected domain D is a K-quasidisk if and only if there exists a constant c such that*

$$h_D(z_1,z_2) \leq c\,j_D(z_1,z_2)$$

for $z_1, z_2 \in D$, where K and c depend only on each other.

We see from Theorem 3.3.12 and Lemma 3.3.5 that D is a quasidisk if and only if the conformally invariant metric h_D is equivalent to the similarity invariant metric j_D in D.

This is also the case if and only if h_D is equivalent to the Möbius invariant metric a_D in D.

THEOREM 3.3.13 (Gehring-Hag [**58**]). *A simply connected domain D is a K-quasidisk if and only if there exists a constant c such that*

$$h_D(z_1,z_2) \leq c\,a_D(z_1,z_2) \tag{3.3.14}$$

for $z_1, z_2 \in D$, where K and c depend only on each other.

We conjecture that a bounded simply connected domain D is a disk if it satisfies (3.3.14) with $c = 1$. This is the case if, in addition, for each $z \in \partial D$ there exists a disk $D' = D'(z) \subset D$ with $z \in \partial D'$. See Gehring-Hag [**56**].

3.4. Geometry of hyperbolic segments

A second useful characterization for quasidisks is an analogue of the following easily established observation.

EXAMPLE 3.4.1. If D is a disk or half-plane and if γ is a hyperbolic segment joining z_1 and z_2 in D, then for each $z \in \gamma$,

$$\text{length}(\gamma) \le \frac{\pi}{2}\,|z_1 - z_2|, \tag{3.4.2}$$

$$\min_{j=1,2} \text{length}(\gamma_j) \le \frac{\pi}{2}\,\text{dist}(z, \partial D), \tag{3.4.3}$$

where γ_1, γ_2 are the components of $\gamma \backslash \{z\}$ and length denotes Euclidean length. The constant $\pi/2$ cannot be replaced by a smaller constant in each of these conclusions.

The situation is different if γ is a hyperbolic geodesic joining z_1 and z_2 in a sector D of angle α. If $0 < \alpha \le \pi$, then (3.4.2) holds as above in Example 3.4.1 while (3.4.3) is replaced by

$$\min_{j=1,2} \text{length}(\gamma_j) \le \frac{\pi}{\alpha}\,\frac{\pi}{2}\,\text{dist}(z, \partial D).$$

On the other hand, if $\pi \le \alpha < 2\pi$, then (3.4.3) holds as in Example 3.4.1 while (3.4.2) is replaced by

$$\text{length}(\gamma) \le \frac{2\pi - \alpha}{\pi}\,\frac{\pi}{2}\,|z_1 - z_2|.$$

These facts lead to the following definition that will allow us to characterize quasidisks in terms of the length and position of their hyperbolic geodesics.

DEFINITION 3.4.4. A simply connected domain $D \subset \mathbf{R}^2$ has the *hyperbolic segment property* if there exists a constant $c \ge 1$ such that for each hyperbolic segment γ joining $z_1, z_2 \in D$ and each $z \in \gamma$,

$$\text{length}(\gamma) \le c\,|z_1 - z_2|,$$
$$\min_{j=1,2} \text{length}(\gamma_j) \le c\,\text{dist}(z, \partial D),$$

where γ_1, γ_2 are the components of $\gamma \setminus \{z\}$.

THEOREM 3.4.5 (Gehring-Osgood [**67**]). *A simply connected domain $D \subset \mathbf{R}^2$ is a K-quasidisk if and only if it has the hyperbolic segment property with constant c, where K and c depend only on each other.*

3.5. Uniform domains

Martio and Sarvas were the first to introduce the surprisingly useful class of *uniform* domains, domains in which each pair of points z_1 and z_2 can be joined by an arc γ with the properties given above in Definition 3.4.4. In particular they point out in Martio-Sarvas [**123**] the important role these domains play in deducing that various classes of locally injective functions are actually injective. The classes include locally bilipschitz functions with small Lipschitz constant and analytic functions with Schwarzian or pre-Schwarzian derivatives which are small relative to the hyperbolic density ρ_D.

DEFINITION 3.5.1. An arbitrary domain $D \subset \mathbf{R}^2$ is *uniform* if there exists a constant $c \geq 1$ such that each pair of points $z_1, z_2 \in D$ can be joined by an arc $\gamma \subset D$ so that for each $z \in \gamma$,

$$\text{length}(\gamma) \leq c\,|z_1 - z_2|, \tag{3.5.2}$$

$$\min_{j=1,2} \text{length}(\gamma_j) \leq c\,\text{dist}(z, \partial D), \tag{3.5.3}$$

where γ_1, γ_2 are the components of $\gamma \setminus \{z\}$.

The hypotheses in Definition 3.4.4 are more restrictive than those above since they require that the arc γ also be a hyperbolic geodesic. On the other hand, the following result implies that when D is simply connected, inequalities (3.5.2) and (3.5.3) hold for a hyperbolic geodesic γ and an absolute constant times c whenever they hold for an arbitrary arc β with constant c and the same endpoints as γ.

LEMMA 3.5.4. *Suppose β is an arc that joins the endpoints z_1, z_2 of a hyperbolic segment γ in a simply connected domain $D \subset \mathbf{R}^2$.*

1° *If*

$$\text{length}(\beta) < c\,|z_1 - z_2|, \tag{3.5.5}$$

then

$$\text{length}(\gamma) \leq ac\,|z_1 - z_2|, \tag{3.5.6}$$

where a is an absolute constant.

2° *If*

$$\min_{j=1,2} \text{length}(\beta_j) \leq c\,\text{dist}(w, \partial D) \tag{3.5.7}$$

for each $w \in \beta$ where β_1, β_2 are the components of $\beta \setminus \{w\}$, then

$$\min_{j=1,2} \text{length}(\gamma_j) \leq b(c+1)\text{dist}(z, \partial D) \tag{3.5.8}$$

for each $z \in \gamma$ where γ_1, γ_2 are the components of $\gamma \setminus \{z\}$ and b is an absolute constant.

PROOF. Since γ is a hyperbolic segment with the same endpoints as β,

$$\text{length}(\gamma) \leq a\,\text{length}(\beta)$$

where a is an absolute constant by Theorem 2 of Gehring-Hayman [**62**]. (See also Theorem 4.20 in Pommerenke [**145**].) Hence (3.5.6) follows from (3.5.5).

If $z \in \gamma$, then by Lemma 2.16 of Gehring-Hag-Martio [**61**], there exists a crosscut α of D which separates the components of $\gamma \setminus \{z\}$ such that

$$\text{length}(\alpha) \leq b_1\,\text{dist}(z, \partial D) \tag{3.5.9}$$

where b_1 is an absolute constant. Hence there exists a point $w \in \beta \cap \alpha$. If the components of $\beta \setminus \{w\}$ and $\gamma \setminus \{z\}$ are labeled so that z_j is the common endpoint of γ_j and β_j, then $\alpha \cup \beta_j$ contains a curve in D which joins the endpoints z_j and z of the hyperbolic segment γ_j. Thus as above,

$$\text{length}(\gamma_j) \leq a\,\text{length}(\alpha \cup \beta_j) \leq a\,\text{length}(\beta_j) + a\,\text{length}(\alpha).$$

Then

$$\min_{j=1,2} \text{length}(\beta_j) \leq c\,\text{dist}(w, \partial D) \leq c\,\text{length}(\alpha)$$

by (3.5.7) and we obtain

$$\min_{j=1,2} \operatorname{length}(\gamma_j) \leq a(c+1) \operatorname{length}(\alpha) \leq b(c+1) \operatorname{dist}(z, \partial D)$$

from (3.5.9) where $b = ab_1$. □

Thus by Theorem 3.4.5, a domain $D \subset \mathbf{R}^2$ is a quasidisk if and only if it is simply connected and uniform.

The following result illustrates the connection between quasidisks and uniform domains for the case where D is multiply connected.

THEOREM 3.5.10 (Gehring [**49**], [**53**], Gehring-Hag [**55**]). *If D is a uniform domain in $\mathbf{R}^2$, then there exists a K such that each component of $\overline{\mathbf{R}}^2 \setminus D$ is either a point or the closure of a K-quasidisk. The converse also holds if D is finitely connected.*

3.6. Min-max property of hyperbolic segments

A second way to characterize quasidisks in terms of the Euclidean properties of their hyperbolic segments is suggested by the following less familiar property of hyperbolic segments in a disk.

EXAMPLE 3.6.1 (Gehring-Hag [**55**]). If D is a disk or half-plane and if γ is a hyperbolic segment joining $z_1, z_2 \in D$, then

$$\frac{1}{\sqrt{2}} \min_{j=1,2} |z_j - w| \leq |z - w| \leq \sqrt{2} \max_{j=1,2} |z_j - w| \tag{3.6.2}$$

for each $z \in \gamma$ and $w \notin D$. The constant $\sqrt{2}$ cannot be replaced by a smaller number in either of these inequalities.

For a proof of this fix $z \in \gamma$, $w \notin D$ and let

$$r_j = \frac{|z_j - w|}{|z - w|}$$

for $j = 1, 2$. We want to show that

$$m = \min_{j=1,2} r_j \leq \sqrt{2}, \qquad M = \max_{j=1,2} r_j \geq \frac{1}{\sqrt{2}}. \tag{3.6.3}$$

Choose a Möbius transformation f which maps z_1, z, z_2 onto $0, 1, \infty$, respectively. Then

$$r_j = \frac{|f(z_j) - f(w)|}{|f(z_j) - f(\infty)|} \frac{|f(\infty) - 1|}{|f(w) - 1|},$$

whence

$$\begin{aligned} |f(w)||f(\infty) - 1| &= r_1 |f(\infty)||f(w) - 1|, \\ |f(\infty) - 1| &= r_2 |f(w) - 1|. \end{aligned} \tag{3.6.4}$$

Since $f(\gamma)$ is a hyperbolic geodesic in $f(D)$, $f(w)$ and $f(\infty)$ lie in the left half-plane,

$$\begin{aligned} 1 + |f(w)|^2 \leq |f(w) - 1|^2 \leq 2\,(1 + |f(w)|^2), \\ 1 + |f(\infty)|^2 \leq |f(\infty) - 1|^2 \leq 2\,(1 + |f(\infty)|^2). \end{aligned} \tag{3.6.5}$$

Then with (3.6.4) and (3.6.5) we obtain

$$\begin{aligned}(1+|f(w)|^2)\,(1+|f(\infty)|^2) &\le (r_1^2|f(\infty)|^2+r_2^2)|f(w)-1|^2\\ &\le 2\,M^2\,(1+|f(w)|^2)\,(1+|f(\infty)|^2)\end{aligned}$$

and

$$\begin{aligned}(1+|f(w)|^2)\,(1+|f(\infty)|^2)\,m^2 &\le (r_1^2|f(\infty)|^2+r_2^2)|f(w)-1|^2\\ &\le 2\,(1+|f(w)|^2)\,(1+|f(\infty)|^2)\end{aligned}$$

from which (3.6.3) follows.

Finally to see that the inequalities in (3.6.2) are sharp for the case where D is a disk, one need only examine the case where the geodesic γ tends to the diameter of D or to a point of ∂D.

Example 3.6.1 leads to a different way of characterizing quasidisks in terms their hyperbolic geodesics.

DEFINITION 3.6.6. A simply connected domain D has the *geodesic min-max property* if there exists a constant $c \ge 1$ such that for each hyperbolic segment γ joining $z_1, z_2 \in D$,

$$\frac{1}{c}\min_{j=1,2}|z_j-w| \le |z-w| \le c\max_{j=1,2}|z_j-w|$$

for each $z \in \gamma$ and $w \notin D$.

THEOREM 3.6.7 (Gehring-Hag [**55**]). *A simply connected domain D is a K-quasidisk if and only if it has the geodesic min-max property with constant c, where K and c depend only on each other.*

3.7. Harmonic measure

The *harmonic measure* of an open or closed arc $\gamma \subset \partial\mathbf{B}$ at a point $z \in \mathbf{B}$ is defined as

$$\omega(z)=\omega(z,\gamma;\mathbf{B})=\int_\gamma P(z,\zeta)\,|d\zeta| \tag{3.7.1}$$

where $P(z,\zeta)$ is the *Poisson kernel*

$$P(z,\zeta)=\frac{1}{2\pi}\,\frac{1-|z|^2}{|z-\zeta|^2}.$$

Thus $\omega(z)$ is the unique function which is bounded and harmonic in $\mathbf{B}$ with boundary values 1 and 0 at interior points of γ and $\partial\mathbf{B}\setminus\overline{\gamma}$, respectively. Note that

$$\omega(0)=\omega(0,\gamma;\mathbf{B})=\frac{\text{length}(\gamma)}{2\pi}.$$

It easily follows that for every conformal self-mapping f of $\mathbf{B}$ we have that

$$\omega(z,\gamma;\mathbf{B})=\omega(f(z),f(\gamma);\mathbf{B}).$$

Suppose next that $D\subset\overline{\mathbf{R}}^2$ is a Jordan domain and that g is a conformal mapping of D onto $\mathbf{B}$. Then g has an extension which maps $\overline{D}$ homeomorphically onto $\overline{\mathbf{B}}$. We define the *harmonic measure* of an arc $\gamma\subset\partial D$ at a point $z\in D$ by

$$\omega(z)=\omega(z,\gamma;D)=\omega(g(z),g(\gamma);\mathbf{B}).$$

By the conformal invariance $\omega(z)$ is independent of the choice of the mapping g.

For example, if D is the right half-plane and γ is the arc in ∂D with endpoints $i\,a, i\,b$ where $0 \le a < b$, then

$$\omega(1, \gamma; D) = \omega(-1, \gamma; D^*) = \frac{1}{\pi}(\arctan b - \arctan a). \tag{3.7.2}$$

For more material on harmonic measure we refer the reader to the book by Garnett and Marshall [**45**].

The formula in (3.7.1) defines the harmonic measure $\omega(z) = \omega(z, C; \mathbf{B})$ for any measurable set $C \subset \partial\mathbf{B}$. With a proper interpretation, harmonic measure is a conformal invariant. It is somewhat surprising that harmonic measure is not a "quasi-invariant" like the modulus of a curve family under quasiconformal mappings (Theorem 1.2.2). Indeed, by a result of Beurling and Ahlfors [**23**] there is a K-quasiconformal mapping $f\colon \mathbf{R}^2 \to \mathbf{R}^2$ with $K < 1 + \epsilon$, for any $\epsilon > 0$, and a compact set $C \subset \partial\mathbf{B}$ such that $f(\mathbf{B}) = \mathbf{B}$ and $\omega(z, C; \mathbf{B}) > 0$ for all $z \in \mathbf{B}$ but $\omega(w, f(C); \mathbf{B}) = 0$ for every $w \in \mathbf{B}$. However, the situation changes if we consider arcs on the boundary of a quasidisk.

We describe two ways of characterizing quasidisks in terms of harmonic measure.

- 1° Harmonic quasisymmetry. Here we compare the harmonic measures of adjacent arcs γ_1 and γ_2 in ∂D at a pair of fixed points z_0 and z_0^* in D and its exterior D^*.
- 2° Harmonic bending. This characterization measures in a conformally invariant way how much each boundary arc $\gamma \subset \partial D$ bends towards the complementary arc $\partial D \setminus \gamma$.

3.8. Harmonic quasisymmetry

This characterization is based on the following symmetry property of a disk or half-plane.

THEOREM 3.8.1 (Hag [**77**]). *A Jordan domain D is a disk or half-plane if and only if there exist points $z_0 \in D$, $z_0^* \in D^*$ such that if γ_1, γ_2 are adjacent arcs in ∂D with*

$$\omega(z_0, \gamma_1; D) = \omega(z_0, \gamma_2; D), \tag{3.8.2}$$

then

$$\omega(z_0^*, \gamma_1; D^*) = \omega(z_0^*, \gamma_2; D^*). \tag{3.8.3}$$

PROOF. Suppose that D is a disk. By performing a preliminary similarity mapping we may assume that $D = \mathbf{B}$. Then $g(z) = z^{-1}$ maps $\mathbf{B}^*$ conformally onto $\mathbf{B}$ and

$$\omega(\infty, \gamma; \mathbf{B}^*) = \omega(0, g(\gamma); \mathbf{B}) = \frac{\text{length}(g(\gamma))}{2\pi} = \frac{\text{length}(\gamma)}{2\pi} = \omega(0, \gamma; \mathbf{B})$$

for each arc $\gamma \subset \partial\mathbf{B}^* = \partial\mathbf{B}$. This yields the desired result with $z_0 = 0$ and $z_0^* = \infty$. The case where D is a half-plane follows similarly.

For the converse choose conformal mappings $f : \mathbf{B} \to D$ and $g : \mathbf{B}^* \to D^*$ and let $f(0) = z_0$ and $g(\infty) = z_0^*$. Then f and g have homeomorphic extensions to $\overline{\mathbf{B}}$ and $\overline{\mathbf{B}^*}$ and, by means of a preliminary rotation, we may assume that $f(1) = g(1)$. The homeomorphism

$$\phi = g^{-1} \circ f : \partial\mathbf{B} \to \partial\mathbf{B}$$

is sense-preserving with $\phi(1) = 1$. We show first that $\phi(z) = z$.

For this let γ_1 and γ_2 be the upper and lower halves of $\partial\mathbf{B}$ labeled so that $i \in \gamma_1$. By the conformal invariance of harmonic measure,

$$\omega(z_0, f(\gamma_1); D) = \omega(0, \gamma_1; \mathbf{B}) = 1/2 = \omega(0, \gamma_2, ; \mathbf{B}) = \omega(z_0, f(\gamma_2); D)$$

and hence by (3.8.2) and (3.8.3)

$$\begin{aligned}\omega(0, \phi(\overline{\gamma}_1); \mathbf{B}) &= \omega(\infty, \phi(\gamma_1); \mathbf{B}^*) = \omega(z_0^*, g \circ \phi(\gamma_1); D^*)\\ &= \omega(z_0^*, f(\gamma_1); D^*) = \omega(z_0^*, f(\gamma_2); D^*)\\ &= \omega(z_0^*, g \circ \phi(\gamma_2); D^*) = \omega(\infty, \phi(\gamma_2); \mathbf{B}^*)\\ &= \omega(0, \phi(\overline{\gamma}_2); \mathbf{B}).\end{aligned}$$

In particular we see that $\phi(-1) = -1$ and, since h is sense-preserving, we see that $\phi(\gamma_1) = \gamma_1$.

Next let γ_1 and γ_2 be the right and left halves of the upper half of $\partial\mathbf{B}$ labeled so that $e^{i\,\pi/4} \in \gamma_1$. Then as above

$$\omega(z_0, f(\gamma_1); D) = \omega(0, \gamma_1; \mathbf{B}) = 1/4 = \omega(0, \gamma_2; \mathbf{B}) = \omega(z_0, f(\gamma_2); D)$$

and

$$\omega(0, \phi(\gamma_1); \mathbf{B}) = \omega(z_0^*, f(\gamma_1); D^*) = \omega(z_0^*, f(\gamma_2); D^*) = \omega(0, \phi(\gamma_2); \mathbf{B}).$$

Thus $\phi(i) = i$ and $\phi(\gamma_1) = \gamma_1$. Proceeding in this way we obtain

$$\phi(e^{2\pi i\,t}) = e^{2\pi i\,t}$$

for all $t \in [0, 1]$ of the form $t = m\,2^{-n}$ where $m, n \in \mathbf{Z}$. Hence by continuity $\phi(z) = z$ for $z \in \partial\mathbf{B}$.

We conclude that f and g together define a self-homeomorphism h of $\overline{\mathbf{R}}^2$ which is conformal in $\mathbf{B} \cup \mathbf{B}^*$ and hence in $\overline{\mathbf{R}}^2$. Thus h is a Möbius transformation and $D = h(\mathbf{B})$ is a disk or half-plane. □

The following counterpart of Theorem 3.8.1 for the case where D is a sector of angle α suggests what the situation is for quasidisks.

EXAMPLE 3.8.4. If $D = \mathbf{S}(\alpha)$ and if γ_1, γ_2 are adjacent arcs in ∂D with

$$\omega(1, \gamma_1; D) = \omega(1, \gamma_2; D), \tag{3.8.5}$$

then

$$\frac{1}{c}\,\omega(-1, \gamma_1; D^*) \le \omega(-1, \gamma_2; D^*) \le c\,\omega(-1, \gamma_1; D^*)$$

where

$$c = \frac{\pi^{1-t}}{2^t - 1}, \qquad t = \min\left(\frac{\alpha}{2\,\pi - \alpha}, \frac{2\,\pi - \alpha}{\alpha}\right).$$

We shall indicate how these bounds are obtained. Suppose that $0 < \alpha \le \pi$ and that $a_1\,e^{i\,\alpha/2}$, $a_2\,e^{i\,\alpha/2}$, $a_3\,e^{i\,\alpha/2}$ are consecutive endpoints of adjacent arcs γ_1 and γ_2 in ∂D for which (3.8.5) holds. Suppose next that $0 \le a_1 < a_2 < a_3$ and set

$$p = \frac{\pi}{\alpha}, \qquad q = \frac{\pi}{2\,\pi - \alpha}, \qquad \frac{\phi}{\pi} = \omega(1, \gamma_j; D)$$

for $j = 1, 2$. Then $f(z) = z^p$ maps D conformally onto the right half-plane $\mathbf{H}$ so that

$$f(1) = 1 \qquad \text{and} \qquad f(a_j\,e^{i\,\alpha/2}) = i\,a_j^p$$

for $j = 1, 2, 3$ and we obtain

$$\omega(1, \gamma_1; D) = \frac{1}{\pi}(\arctan a_2^p - \arctan a_1^p) = \frac{\phi}{\pi},$$
$$\omega(1, \gamma_2; D) = \frac{1}{\pi}(\arctan a_3^p - \arctan a_2^p) = \frac{\phi}{\pi}$$

from (3.7.2) and the conformal invariance of harmonic measure. A similar argument yields

$$\omega(-1, \gamma_1; D^*) = \frac{1}{\pi}(\arctan a_2^q - \arctan a_1^q),$$
$$\omega(-1, \gamma_2; D^*) = \frac{1}{\pi}(\arctan a_3^q - \arctan a_2^q)$$

and we conclude that

$$\frac{\omega(-1, \gamma_2; D^*)}{\omega(-1, \gamma_1; D^*)} = \frac{g(\theta + \phi, t) - g(\theta, t)}{g(\theta, t) - g(\theta - \phi, t)}$$

where

$$g(\theta, t) = \arctan(\tan^t \theta) \qquad \text{and} \qquad t = \frac{q}{p} = \frac{\alpha}{2\pi - \alpha}. \tag{3.8.6}$$

A technical but elementary argument then shows that

$$\frac{1}{c} \leq \frac{g(\theta + \phi, t) - g(\theta, t)}{g(\theta, t) - g(\theta - \phi, t)} \leq 1$$

for $0 \leq \theta - \phi < \theta + \phi \leq \pi/2$ and this yields the desired conclusion if γ_1 and γ_2 lie on the same ray of ∂D. The general conclusion can now be deduced from this special case.

The preceding example suggests the following quasisymmetry property. See Krzyż [**105**].

DEFINITION 3.8.7. A Jordan domain D is *quasisymmetric* if there exist points $z_0 \in D$, $z_0^* \in D^*$ and a constant $c \geq 1$ such that if γ_1, γ_2 are adjacent arcs in ∂D with

$$\omega(z_0, \gamma_1; D) = \omega(z_0, \gamma_2; D),$$

then

$$\frac{1}{c}\,\omega(z_0^*, \gamma_1; D^*) \leq \omega(z_0^*, \gamma_2; D^*) \leq c\,\omega(z_0^*, \gamma_1; D^*).$$

THEOREM 3.8.8 (Krzyż [**105**]). *A Jordan domain D is a K-quasidisk if and only if it is quasisymmetric with constant c, where K and c depend only on each other.*

This result will follow from the discussion in Section 3.12 where it is proved that a domain is quasisymmetric if and only if it satisfies the conjugate quadrilateral inequality. See Definition 3.10.8 below.

3.9. Harmonic bending

Suppose that γ is a closed arc with endpoints z_1 and z_2. Then $D = \overline{\mathbf{R}}^2 \setminus \gamma$ is a simply connected domain and there exists a conformal mapping g which maps D onto the right half-plane $\mathbf{H}$. Next since D is locally connected at z_1 and z_2, g has a continuous injective extension in $D \cup \{z_1, z_2\}$ (Väisälä [**163**]). Hence we may choose g so that $g(z_1) = 0$ and $g(z_2) = \infty$. For $z \in D$ let

$$b(z, \gamma) = \max_{j=1,2} \omega(g(z), \gamma_j; \mathbf{H}),$$

where γ_1 and γ_2 are the positive and negative halves of the imaginary axis. Then

$$b(z, \gamma) \to 1 \qquad \text{if and only if} \qquad z \to \gamma \setminus \{z_1, z_2\}.$$

Hence the function $b(z, \gamma)$ is a conformally invariant measure of the position of the point $z \in D$ with respect to the interior of γ which attains its minimum $1/2$ on the preimage of the real axis under g.

If γ is a closed subarc of a Jordan curve $C \subset \overline{\mathbf{R}}^2$, then the function $b(z, \gamma)$ measures how much the arc $C \setminus \gamma$ bends towards γ. In particular, if D is a disk or half-plane, then for each closed arc $\gamma \subset \partial D$,

$$b(z, \gamma) = \frac{1}{2}$$

for $z \in \partial D \setminus \gamma$.

The following observation suggests how the bending function $b(z, \gamma)$ may be used to characterize quasidisks.

EXAMPLE 3.9.1. If D is a sector of angle α and if γ is an arc in ∂D, then

$$b(z, \gamma) \leq \max(\frac{\alpha}{2\pi}, 1 - \frac{\alpha}{2\pi})$$

for each $z \in \partial D \setminus \gamma$. The above bound is sharp.

DEFINITION 3.9.2. A Jordan domain D has the *harmonic bending property* if there exists a constant $c \in [\frac{1}{2}, 1)$ such that for each closed arc $\gamma \subset \partial D$,

$$b(z, \gamma) \leq c$$

for $z \in \partial D \setminus \gamma$.

THEOREM 3.9.3 (Fernández-Hamilton-Heinonen [**41**]). *A Jordan domain D is a K-quasidisk if and only if it has the harmonic bending property with constant c, where K and c depend only on each other.*

A Jordan domain D is a disk or half-plane if and only if it has the harmonic bending property with constant $c = \frac{1}{2}$.

3.10. Quadrilaterals

A *quadrilateral* $Q = D(z_1, z_2, z_3, z_4)$ consists of a Jordan domain $D \subset \overline{\mathbf{R}}^2$ together with a quadruple of points $z_1, z_2, z_3, z_4 \in \partial D$, the *vertices* of Q, that are positively oriented with respect to D. The vertices z_1, z_2, z_3, z_4 divide ∂D into four arcs, the *sides* of Q.

Each such quadrilateral Q can be mapped conformally onto a rectangle $R = R(0, a, a+i, i)$ so that the vertices and sides of Q and R correspond. The *modulus* of Q is then given by

$$\text{mod}(Q) = a.$$

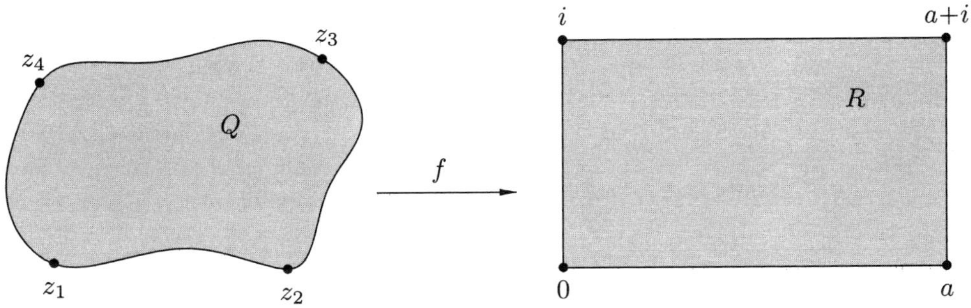

FIGURE 3.2

The modulus of a quadrilateral Q can also be given in terms of the modulus of the family of curves joining opposite sides of Q.

LEMMA 3.10.1. *If $Q = D(z_1, z_2, z_3, z_4)$ is a quadrilateral, then*

$$\text{mod}(Q) = \text{mod}(\Gamma) \tag{3.10.2}$$

where Γ is the family of arcs in Q which join the sides of Q with endpoints z_1, z_2 and z_3, z_4.

PROOF. Since $\text{mod}(Q)$ and $\text{mod}(\Gamma)$ are conformally invariant, it suffices to consider the case where Q is the rectangle $R = R(0, a, a+i, i)$. The desired conclusion then follows from Lemma 1.3.1. □

Recall that the cross ratio of four points z_1, z_2, z_3, z_4 in $\overline{\mathbf{R}}^2$ is denoted by $[z_1, z_2, z_3, z_4]$. By choosing $a = 1$ in the rectangle $R = R(0, a, a+i, i)$ we immediately have the following corollary to Lemma 3.10.1 whenever D is a disk or a half-plane.

COROLLARY 3.10.3. *Let Q be the quadrilateral $\mathbf{B}(z_1, z_2, z_3, z_4)$ or $\mathbf{H}(z_1, z_2, z_3, z_4)$. Then we have that*

$$\text{mod}(Q) = 1$$

if and only if

$$[z_1, z_2, z_3, z_4] = 2.$$

The next result yields a more general relation between the modulus of a quadrilateral and the cross ratio of its vertices in the disk or half-plane setting.

LEMMA 3.10.4. *Suppose that Q is the quadrilateral $\mathbf{B}(z_1, z_2, z_3, z_4)$ or $\mathbf{H}(z_1, z_2, z_3, z_4)$, where $z_1 < z_2 < z_3 < z_4$ if $D = \mathbf{H}$. Then $1 < [z_1, z_2, z_3, z_4] < \infty$ and*

$$\text{mod}(Q) = \frac{2}{\pi}\,\mu\left(\frac{1}{\sqrt{[z_1, z_2, z_3, z_4]}}\right) \tag{3.10.5}$$

where

$$\mu(r) = \frac{\pi}{2}\,\frac{\mathcal{K}(\sqrt{1-r^2})}{\mathcal{K}(r)} \tag{3.10.6}$$

with $\mathcal{K}(r)$ the complete elliptic integral of the first kind,

$$\mathcal{K}(r) = \int_0^1 \frac{dt}{\sqrt{(1-t^2)(1-r^2t^2)}},$$

a strictly decreasing homeomorphism of $(0,1)$ *onto* $(0,\infty)$. *Furthermore, we have that*

$$\mu(r) \simeq \log\frac{4}{r}$$

as $r \to 0$ *in* (3.10.6).

SKETCH OF PROOF. Let $f : \mathbf{H} \to \mathbf{H}$ be a Möbius transformation which maps z_1, z_2, z_3, z_4 onto $-1/r, -1, 1, 1/r$, respectively, where $0 < r < 1$. Then

$$[z_1, z_2, z_3, z_4] = \left(\frac{1+r}{2\sqrt{r}}\right)^2$$

and the transformation

$$g(w) = \int_0^w \frac{d\omega}{\sqrt{(1-\omega^2)(1-r^2\,\omega^2)}}$$

maps the quadrilateral $Q' = \mathbf{H}(-1/r, -1, 1, 1/r)$ conformally onto the quadrilateral $Q'' = R(-K+iK', -K, K, K+iK')$ so that vertices correspond, where

$$K = \mathcal{K}(r) \qquad \text{and} \qquad K' = \mathcal{K}(\sqrt{1-r^2}).$$

Then

$$\begin{aligned}\operatorname{mod}(Q) = \operatorname{mod}(Q') = \operatorname{mod}(Q'') = \frac{K'}{2K} = \frac{1}{\pi}\,\mu(r) = \frac{2}{\pi}\,\mu\left(\frac{2\sqrt{r}}{1+r}\right) \\ = \frac{2}{\pi}\,\mu\left(\frac{1}{\sqrt{[z_1, z_2, z_3, z_4]}}\right).\end{aligned}$$

See Anderson-Vamanamurthy-Vuorinen [**11**], [**12**] and Lehto-Virtanen [**117**]. □

Suppose next that C is a Jordan curve which bounds the domains D and D^* and that $z_1, z_2, z_3, z_4 \in C$ is a quadruple of points positively oriented with respect to D. Then the quadrilaterals $Q = D(z_1, z_2, z_3, z_4)$ and $Q^* = D^*(z_4, z_3, z_2, z_1)$ are said to be *conjugate* with respect to C.

In particular if D is a disk or half-plane and if Q and Q^* are quadrilaterals conjugate with respect to ∂D, then

$$\operatorname{mod}(Q^*) = \operatorname{mod}(Q).$$

The following example shows what to expect when D is a quasidisk.

EXAMPLE 3.10.7. If D is a sector of angle α and if Q and Q^* are quadrilaterals conjugate with respect to ∂D, then

$$\min\left(\frac{2\pi-\alpha}{\alpha}, \frac{\alpha}{2\pi-\alpha}\right) \le \frac{\operatorname{mod}(Q^*)}{\operatorname{mod}(Q)} \le \max\left(\frac{2\pi-\alpha}{\alpha}, \frac{\alpha}{2\pi-\alpha}\right)$$

and these bounds are sharp.

To see this, suppose that $D = \mathbf{S}(\alpha)$. By the discussion in Example 1.4.2 there exists a K-quasiconformal self-mapping of $\mathbf{R}^2$ which maps $\mathbf{S}(\alpha)$ onto $\mathbf{S}^*(\alpha)$ where

$$K = \max\left(\frac{2\pi-\alpha}{\alpha}, \frac{\alpha}{2\pi-\alpha}\right).$$

Hence the above bounds follow from Theorem 1.2.2 and Lemma 3.10.1.

To show that these bounds are sharp, suppose that $0 < \alpha \leq \pi$ and for $1 < r < \infty$ let $Q = D(z_1, z_2, z_3, z_4)$ denote the quadrilateral in $D = \mathbf{S}(\alpha)$ with vertices $z_1 = r\, e^{i\,\alpha/2}$, $z_2 = e^{i\,\alpha/2}$, $z_3 = e^{-i\,\alpha/2}$, $z_4 = r\, e^{-i\,\alpha/2}$ and set

$$p = \frac{\pi}{\alpha}, \qquad q = \frac{\pi}{2\,\pi - \alpha}.$$

Then $f(z) = i\, z^p$ maps Q conformally onto $Q' = \mathbf{H}(w_1, w_2, w_3, w_4)$ where $w_1 = -r^p$, $w_2 = -1$, $w_3 = 1$, $w_4 = r^p$. Hence by Lemma 3.10.4

$$\operatorname{mod}(Q) = \operatorname{mod}(Q') = \frac{2}{\pi}\, \mu\left(\frac{2\, r^{-p/2}}{1 + r^{-p}}\right) \simeq \frac{p}{\pi} \log r$$

as $r \to \infty$. Similarly

$$\operatorname{mod}(Q^*) = \frac{2}{\pi}\, \mu\left(\frac{2\, r^{-q/2}}{1 + r^{-q}}\right) \simeq \frac{q}{\pi} \log r$$

as $r \to \infty$,

$$\lim_{r \to \infty} \frac{\operatorname{mod}(Q^*)}{\operatorname{mod}(Q)} = \frac{2\,\pi - \alpha}{\alpha},$$

and hence the upper bound in Example 3.10.7 is sharp. A similar argument yields the same result for the lower bound.

If D is a sector of angle α, then Example 3.10.7 implies that

$$\operatorname{mod}(Q^*) \leq \max\left(\frac{2\,\pi - \alpha}{\alpha}, \frac{\alpha}{2\,\pi - \alpha}\right)$$

whenever Q^* is a quadrilateral conjugate to Q with respect to ∂D with $\operatorname{mod}(Q) = 1$. We are thus led to one of the first published criteria for quasidisks (Tienari [**159**], Lehto-Virtanen [**117**], Pfluger [**144**]).

DEFINITION 3.10.8. A Jordan domain D satisfies the *conjugate quadrilateral inequality* if there exists a constant $c \geq 1$ such that if Q and Q^* are conjugate quadrilaterals with respect to ∂D and if $\operatorname{mod}(Q) = 1$, then

$$\operatorname{mod}(Q^*) \leq c.$$

The following example shows that the above property characterizes disks and half-planes when $c = 1$.

EXAMPLE 3.10.9. A Jordan domain D is a disk or half-plane if and only if it satisfies the conjugate quadrilateral inequality with $c = 1$.

To establish the sufficiency, suppose that f and g map D and D^* conformally onto the upper and lower half-planes $\mathbf{H}$ and $\mathbf{H}^*$, respectively. Then f and g have homeomorphic extensions to $\overline{D}$ and $\overline{D}^*$ and $\phi = f \circ g^{-1}$ is a self-homeomorphism of $\overline{\mathbf{R}}^1$. By renormalizing we may assume that ϕ fixes the points $0, 1, \infty$.

The quadrilaterals $Q = \mathbf{H}(a, b, c, \infty)$ and $Q^* = \mathbf{H}^*(\phi(a), \phi(b), \phi(c), \infty)$ have modulus 1 if and only if

$$b - a = c - b \qquad \text{and} \qquad \phi(b) - \phi(a) = \phi(c) - \phi(b),$$

respectively. By hypothesis $\operatorname{mod}(Q^*) = 1$ whenever $\operatorname{mod}(Q) = 1$. It then follows by induction that $\phi(x) = x$, first for $x = m$ where $m = 0, \pm 1, \pm 2, \ldots$ and then for

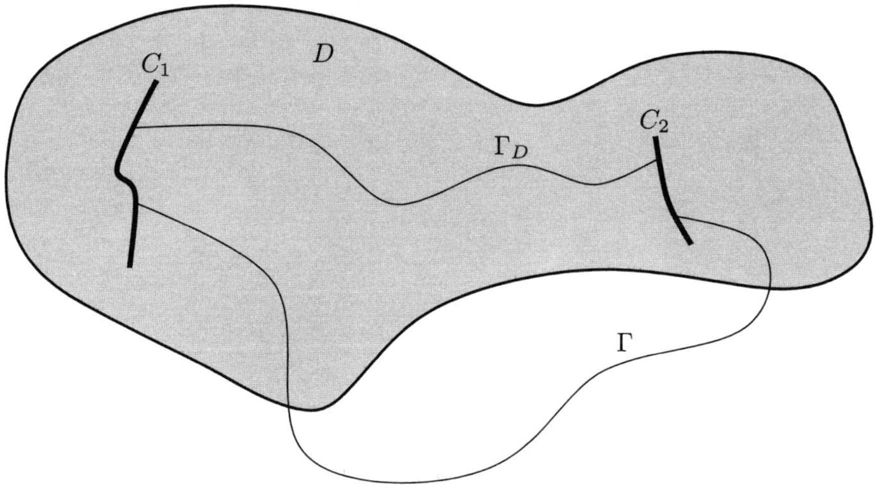

FIGURE 3.3

$x = m\,2^{-n}$ where $n = 0, 1, 2, \ldots$. Hence $\phi(x) = x$ for all $x \in \mathbf{R}^1$ by continuity and

$$h(z) = \begin{cases} f(z) & \text{if } z \in \overline{D}, \\ g(z) & \text{if } z \in D^* \end{cases}$$

defines a Möbius transformation which maps D onto $\mathbf{H}$.

THEOREM 3.10.10 (Tienari [**159**], Lehto-Virtanen [**117**], Pfluger [**144**]). *A Jordan domain D is a K-quasidisk if and only if it satisfies the conjugate quadrilateral inequality with constant c, where K and c depend only on each other.*

3.11. Extremal distance property

We observed earlier that $\text{mod}(\Gamma)$ is a conformally invariant outer measure of a family of curves which is large if the curves in Γ are short and plentiful and which is small if the curves are long or scarce. We now use this quantity to compare distances between two continua $C_1, C_2 \subset D$, as measured by the moduli of the families of curves which join them in D and in $\overline{\mathbf{R}}^2$, respectively. This, in turn, leads to another characterization for quasidisks.

Given continua $C_1, C_2 \subset D$, we denote by Γ_D and Γ the families of all curves which join C_1 and C_2 in D and $\overline{\mathbf{R}}^2$, respectively. We find it convenient to introduce the notation

$$\mu_D(C_1, C_2) = \text{mod}(\Gamma_D) \qquad \text{and} \qquad \mu(C_1, C_2) = \text{mod}(\Gamma).$$

Since $\Gamma_D \subset \Gamma$,

$$\mu_D(C_1, C_2) \leq \mu(C_1, C_2)$$

for all domains D. On the other hand, for certain domains D we can, up to a constant, reverse the direction of the above inequality.

EXAMPLE 3.11.1 (Yang [**169**]). If D is a sector of angle α, then

$$\mu(C_1, C_2) \leq \frac{2\,\pi}{\alpha}\,\mu_D(C_1, C_2) \tag{3.11.2}$$

for each pair of continua C_1 and C_2 in D. The bound in (3.11.2) is sharp.

Example 3.11.1 suggests another characteristic property for the class of quasidisks.

DEFINITION 3.11.3. A domain D has the *extremal distance property* if there exists a constant c such that

$$\mu(C_1, C_2) \leq c\, \mu_D(C_1, C_2) \tag{3.11.4}$$

for all continua $C_1, C_2 \subset D$.

The existence of such a constant c implies that D is not bent around part of its exterior D^* so that the Euclidean distance between C_1 and C_2 in D is substantially larger than the distance in $\overline{\mathbf{R}}^2$.

THEOREM 3.11.5 (Gehring-Martio [**65**]). *A simply connected domain D is a K-quasidisk if and only if it has the extremal distance property with constant c, where K and c depend only on each other.*

The constant c in the extremal distance property for a simply connected domain D is never less than 2. Moreover D is a disk or half-plane whenever $c = 2$ (Yang [**168**]).

The following is an attractive application of the above characterization for quasidisks.

THEOREM 3.11.6 (Fernández-Heinonen-Martio [**42**]). *Suppose that $f : D \to D'$ is a conformal mapping and that D' is a quasidisk. If $E \subset D$ is a quasidisk, then so is $E' = f(E)$.*

PROOF. Choose continua $C_1', C_2' \subset E'$, let $C_j = f^{-1}(C_j')$ for $j = 1, 2$, and let c and c' be the respective extremal distance constants for E and D'. Then by Theorem 3.11.5 and the conformal invariance of extremal distance,

$$\begin{aligned} \mu(C_1', C_2') &\leq c'\, \mu_{D'}(C_1', C_2') = c'\, \mu_D(C_1, C_2) \leq c'\, \mu(C_1, C_2) \\ &\leq cc'\, \mu_E(C_1, C_2) = cc'\, \mu_{E'}(C_1', C_2'). \end{aligned}$$

Thus E' has the extremal distance property with constant cc' and hence is a quasidisk. □

Theorem 3.11.6 is a variant of the so-called *subinvariance principle.* Suppose that $f : D \to D'$ is conformal where D' is a disk. According to this principle, if $E \subset D$ is a *nice* set, then so is $E' = f(E)$. Here are two examples.

The first is a special case of what was proved above.

EXAMPLE 3.11.7. If E is a quasidisk in D, then so is E'.

The second example concerns the length of a *linear* set, i.e., a subset of a line. If E is a linear set in D, then a simple projection argument shows that

$$\text{length}(E) \leq \frac{1}{2}\, \text{length}(\partial D).$$

We get a similar conclusion for the image E' of E.

EXAMPLE 3.11.8. If E in a linear set in D, then

$$\text{length}(E') \leq 2\, \text{length}(\partial D'). \tag{3.11.9}$$

Example 3.11.8 is a reformulation of the *level-set inequality*, first established by Hayman and Wu [**79**] with an absolute constant c in place of 2 in (3.11.9). Øyma [**142, 143**] later showed that (3.11.9) holds with the constant 2, and Rohde [**152**] proved that in fact a strict inequality holds in (3.11.9). See also Fernández-Heinonen-Martio [**42**], Flinn [**43**], Garnett-Gehring-Jones[**46**], Garnett-Marshall [**45**].

3.12. Quadrilaterals and harmonic quasisymmetry

The two characterizations for a quasidisk D involving the conjugate quadrilateral inequality and harmonic quasisymmetry compare the conformal geometry of a Jordan domain D with that of its exterior D^*. In particular they compare the moduli of the quadrilaterals in D and D^* determined by an ordered quadruple of points on their common boundary ∂D and the harmonic measures of two adjacent arcs in ∂D evaluated at interior points in D and D^*.

We conclude this chapter by showing that these two characterizations are equivalent. We do this by replacing the two Jordan domains D and D^* by a single domain, the unit disk $\mathbf{B}$, the pair of interior points in D and D^* by the origin $0 \in \mathbf{B}$, and the common boundary $\partial D = \partial D^*$ by a self-homeomorphism h of $\partial \mathbf{B}$. More specifically we choose conformal mappings $f : D \to \mathbf{B}$ and $g : D^* \to \mathbf{B}$ and let h denote the induced *sewing* homeomorphism

$$h = g \circ f^{-1} : \partial \mathbf{B} \to \partial \mathbf{B}. \tag{3.12.1}$$

As we shall see, relations between D and D^* are then encoded in properties of the sewing homeomorphism h. For this we look at the cross ratios and quasisymmetry of points on the unit circle $\partial \mathbf{B}$. In particular we assume in what follows that ϕ is a self-homeomorphism of $\partial \mathbf{B}$ and that z_1, z_2, z_3, z_4 is an ordered quadruple of points in $\partial \mathbf{B}$.

We establish two lemmas concerning cross ratios which appear in a more general setting in a paper by Väisälä [**164**].

LEMMA 3.12.2. *Suppose that ϕ fixes the points $i, -1, -i$ and that a is a constant with $a \geq 2$. If*

$$\frac{a}{a-1} \leq \frac{|\phi(z_1) - \phi(z_3)|}{|\phi(z_1) - \phi(z_4)|} \frac{|\phi(z_2) - \phi(z_4)|}{|\phi(z_2) - \phi(z_3)|} \leq a \tag{3.12.3}$$

whenever $[z_1, z_2, z_3, z_4] = 2$, then

$$\frac{1}{b} \leq \frac{|\phi(z_1) - \phi(z_2)|}{|\phi(z_2) - \phi(z_3)|} \leq b \tag{3.12.4}$$

whenever $|z_1 - z_2| = |z_2 - z_3|$ where $b = a^2 + 1$.

PROOF. We begin by determining a lower bound for $|\phi(1) \pm i|$ that will be needed later. Since $[1, i, -1, -i] = [i, -1, -i, 1] = 2$,

$$\frac{|\phi(1) + 1|}{|\phi(1) + i|} \frac{|i + i|}{|i + 1|} \leq a, \qquad \frac{|i + i|}{|i - \phi(1)|} \frac{|1 + \phi(1)|}{|1 - i|} \leq a$$

by (3.12.3) and we obtain

$$|\phi(1) \pm i| \geq \frac{2}{a} \tag{3.12.5}$$

since $|\phi(1) + 1| > \sqrt{2}$.

Now suppose that z_1, z_2, z_3 is a triple of points in $\partial\mathbf{B}$ with

$$|z_1 - z_2| = |z_2 - z_3|$$

and choose $z_4 = -z_2$. Then z_1, z_2, z_3, z_4 is an ordered quadruple of points in $\partial\mathbf{B}$ with $[z_1, z_2, z_3, z_4] = 2$, by elementary geometry.

We claim that

$$|\phi(z_2) - \phi(z_4)| \geq \frac{2}{a}. \tag{3.12.6}$$

By symmetry it is sufficient to study the cases $\arg(z_2) \in [0, \pi/2]$ and $\arg(z_2) \in [\pi/2, \pi]$. In the first case $\arg(z_4) \in [\pi, 3\pi/2]$ and $\arg(\phi(z_4)) \in [\pi, 3\pi/2]$. Moreover, $\arg(\phi(z_2)) \in [-\pi/2, \pi/2]$. Since $|\phi(1) + i| \geq 2/a$ and ϕ is order-preserving, we have that $|\phi(z_4) + i| \geq 2/a$. This implies that $|\phi(z_2) - \phi(z_4)| \geq \min\{2/a, \sqrt{2}\} = 2/a$. The second case is proved in a similar way.

Next (3.12.6) implies that

$$\frac{2/a}{2} \frac{|\phi(z_1) - \phi(z_3)|}{|\phi(z_2) - \phi(z_3)|} \leq \frac{|\phi(z_1) - \phi(z_3)|}{|\phi(z_1) - \phi(z_4)|} \frac{|\phi(z_2) - \phi(z_4)|}{|\phi(z_2) - \phi(z_3)|} \leq a,$$

whence

$$|\phi(z_1) - \phi(z_2)| \leq |\phi(z_1) - \phi(z_3)| + |\phi(z_2) - \phi(z_3)| \leq b\,|\phi(z_2) - \phi(z_3)|$$

where $b = a^2 + 1$. This is the upper bound in (3.12.4).

Finally from Ptolemy's identity

$$|z_1 - z_2||z_3 - z_4| + |z_2 - z_3||z_4 - z_1| = |z_1 - z_3||z_2 - z_4| \tag{3.12.7}$$

we see that z_3, z_2, z_1, z_4 is a quadruple of points in $\partial\mathbf{B}$ and with $[z_3, z_2, z_1, z_4] = 2$. Hence we can interchange the points z_1 and z_3 in the above argument to obtain the lower bound in (3.12.4). □

We consider next a converse of Lemma 3.12.2. Our proof depends on the following observation concerning the geometry of quadrilaterals Q in $\mathbf{B}$ with $\text{mod}(Q) = 1$.

REMARK 3.12.8. If z_1, z_2, z_3, z_4 is an ordered quadruple of points in $\partial\mathbf{B}$ with

$$|z_1 - z_2||z_3 - z_4| = |z_2 - z_3||z_4 - z_1|,$$
$$|z_1 - z_2| \leq \min(|z_2 - z_3|, |z_4 - z_1|),$$

then

$$|z_3 - z_4| < (\sqrt{2} + 1)\max(|z_2 - z_3|, |z_4 - z_1|),$$
$$|z_1 - z_2| > (\sqrt{2} - 1)\min(|z_2 - z_3|, |z_4 - z_1|).$$

Both bounds are best possible.

PROOF. Let

$$a = |z_1 - z_2|, \qquad b = |z_2 - z_3|, \qquad c = |z_3 - z_4|, \qquad d = |z_4 - z_1|.$$

Then by hypothesis,

$$ac = bd, \qquad a \leq \min(b, d) \leq \max(b, d) \leq c.$$

By interchanging z_1, z_2 and z_3, z_4 if necessary, we may also assume that

$$a \leq b \leq d \leq c.$$

The square of the length l of the diagonal (z_1, z_3) is then given by

$$l^2 = c^2 + d^2 - 2cd\cos\theta = a^2 + b^2 + 2ab\cos\theta,$$

where θ is the angle formed by the segments (z_4, z_1) and (z_4, z_3), and we obtain

$$c^2 - a^2 \le c^2 - a^2 + d^2 - b^2 = 2(cd + ab)\cos\theta \le 2(c+a)d\cos\theta.$$

Thus

$$0 \le c - a \le 2d\cos\theta \qquad \text{and} \qquad \frac{c}{d} - \frac{a}{b} \le \frac{c}{d} - \frac{a}{d} \le 2\cos\theta.$$

Hence

$$x - 2\cos\theta \le \frac{1}{x}, \qquad x \le \sqrt{1+\cos^2\theta} + \cos\theta$$

where $x = c/d$, and we obtain

$$\frac{c}{d} = x < \sqrt{2} + 1, \qquad \frac{a}{b} = \frac{1}{x} > \sqrt{2} - 1.$$

Letting $\theta \to 0$ in the above calculations shows that the constants $\sqrt{2}+1$ and $\sqrt{2}-1$ cannot be improved. □

LEMMA 3.12.9. *Suppose b is a constant with $b \ge 1$. If*

$$\frac{1}{b} \le \frac{|\phi(z_1) - \phi(z_2)|}{|\phi(z_2) - \phi(z_3)|} \le b \tag{3.12.10}$$

whenever $|z_1 - z_2| = |z_2 - z_3|$, then

$$\frac{a}{a-1} \le \frac{|\phi(z_1) - \phi(z_3)|\,|\phi(z_2) - \phi(z_4)|}{|\phi(z_1) - \phi(z_4)|\,|\phi(z_2) - \phi(z_3)|} \le a \tag{3.12.11}$$

whenever $[z_1, z_2, z_3, z_4] = 2$ where $a = a(b) \ge 2$.

PROOF. We begin by showing that if z_i, z_j, z_k is a triple of points in $\partial\mathbf{B}$ with

$$|z_j - z_k| \le |z_i - z_j|, \tag{3.12.12}$$

then

$$|\phi(z_i) - \phi(z_k)| \le c\,|\phi(z_i) - \phi(z_j)|, \qquad c = b + 1. \tag{3.12.13}$$

For this let γ denote the arc in $\partial\mathbf{B}$ with endpoints z_i, z_k which contains z_j and let ζ denote the midpoint of γ. Next we may assume that $\phi(z_k)$ is on the larger arc of $\partial\mathbf{B}$ connecting $\phi(z_i)$ and $\phi(z_j)$ since otherwise (3.12.13) follows trivially. Then $\phi(\zeta)$ lies on the smaller arc connecting $\phi(z_i)$ and $\phi(z_j)$. Hence

$$\begin{aligned}|\phi(z_i) - \phi(z_k)| &\le |\phi(z_i) - \phi(\zeta)| + |\phi(\zeta) - \phi(z_k)| \le c\,|\phi(z_i) - \phi(\zeta)|\\ &\le c\,|\phi(z_i) - \phi(z_j)|\end{aligned}$$

by (3.12.10).

Suppose now that z_1, z_2, z_3, z_4 is an ordered quadruple of points in $\partial\mathbf{B}$ with $[z_1, z_2, z_3, z_4] = 2$. Then

$$\begin{aligned}|z_1 - z_2||z_3 - z_4| &= |z_1 - z_3||z_2 - z_4| - |z_2 - z_3||z_4 - z_1|\\ &= [z_1, z_2, z_3, z_4]\,|z_2 - z_3||z_4 - z_1| - |z_2 - z_3||z_4 - z_1|\\ &= |z_2 - z_3||z_4 - z_1|\end{aligned}$$

by (3.12.7), and by interchanging the pairs of indices $(1,2)$ and $(3,4)$ and the pairs $(2,3)$ and $(1,4)$ when appropriate, we need only consider the following two cases:

$$1^\circ\ |z_3 - z_4| \le |z_2 - z_3|, |z_4 - z_1|,$$
$$2^\circ\ |z_4 - z_1| \le |z_1 - z_2| \le |z_3 - z_4| \le |z_2 - z_3|.$$

In the first case we see that

$$|\phi(z_1) - \phi(z_3)| \le c\,|\phi(z_1) - \phi(z_4)|, \qquad |\phi(z_2) - \phi(z_4)| \le c\,|\phi(z_2) - \phi(z_3)|$$

by (3.12.13) applied to the triples of points $\{z_1, z_4, z_3\}$, $\{z_2, z_3, z_4\}$. This, in turn, yields the upper bound in (3.12.11) with $a = (b+1)^2$.

In the second case we obtain

$$|\phi(z_1) - \phi(z_3)| \le c\,|\phi(z_2) - \phi(z_3)|$$

from (3.12.13) applied to the triple $\{z_3, z_2, z_1\}$. Next we choose z_0 on the arc of $\partial\mathbf{B}$ with endpoints z_1, z_2 so that

$$|z_0 - z_1| = |z_1 - z_4| > (\sqrt{2} - 1)|z_1 - z_2|,$$

by Remark 3.12.8. Then a numerical computation shows that $|z_2 - z_0| < |z_0 - z_4|$ and we obtain

$$|\phi(z_2) - \phi(z_4)| \le c|\phi(z_4) - \phi(z_0)| \le c^2\,|\phi(z_4) - \phi(z_1)|.$$

This pair of bounds yields the upper bound in (3.12.11) with $a = (b+1)^3$.

To obtain the lower bound in (3.12.11), we observe that $[z_1, z_2, z_3, z_4] = 2$ implies $[z_1, z_4, z_3, z_2] = 2$ by (3.12.7). Likewise $[\phi(z_1), \phi(z_2), \phi(z_3), \phi(z_4)]^{-1} = 1 - [\phi(z_1), \phi(z_4), \phi(z_3), \phi(z_2)]^{-1}$ and the lower bound follows from the upper bound. □

We now apply Lemmas 3.12.2 and 3.12.9 to obtain the main result of this section.

THEOREM 3.12.14 (Gehring-Hag [**59**]). *Suppose that D is a Jordan domain in $\overline{\mathbf{R}}^2$. Then D is quasisymmetric with respect to harmonic measure if and only if D satisfies the conjugate quadrilateral inequality.*

PROOF. Suppose that D satisfies the conjugate quadrilateral inequality and choose conformal mappings $f : D \to \mathbf{B}$ and $g : D^* \to \mathbf{B}$ normalized so that ϕ fixes the points $i, -1, -i$ where

$$\phi = \overline{h}, \qquad h = g \circ f^{-1} : \partial\mathbf{B} \to \partial\mathbf{B}.$$

We shall show that D satisfies the harmonic quasisymmetry condition with respect to the points $w_0 = f^{-1}(0) \in D$ and $w_0^* = g^{-1}(0) \in D^*$.

For this suppose that $Q = D(w_1, w_2, w_3, w_4)$ is a quadrilateral with $\operatorname{mod}(Q) = 1$. Then by hypothesis, $\operatorname{mod}(Q^*) \le c$. The corresponding quadrilaterals $f(Q)$ and $g(Q^*)$ in $\mathbf{B}$ are $\mathbf{B}(z_1, z_2, z_3, z_4)$ and $\mathbf{B}(h(z_4), h(z_3), h(z_2), h(z_1))$ and (3.10.5) and (3.10.6) imply there exists a constant $a = a(c) \ge 2$ such that

$$\frac{a-1}{a} \le \frac{|h(z_1) - h(z_3)|\,|h(z_2) - h(z_4)|}{|h(z_1) - h(z_4)|\,|h(z_2) - h(z_3)|} \le a. \tag{3.12.15}$$

Then Lemma 3.12.2 with $\phi = \overline{h}$ implies that

$$\frac{1}{b} \le \frac{|h(z_1) - h(z_2)|}{|h(z_2) - h(z_3)|} \le b, \tag{3.12.16}$$

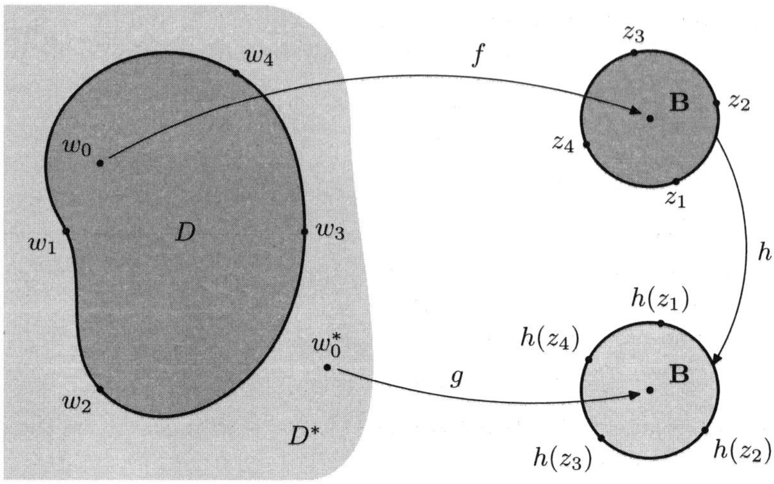

FIGURE 3.4

whenever $|z_1 - z_2| = |z_2 - z_3|$, i.e., whenever z_1, z_2 and z_2, z_3 are the endpoints of a pair of adjacent open arcs γ_1 and γ_2 in $\partial\mathbf{B}$ with

$$\omega(0, \gamma_1; \mathbf{B}) = \omega(0, \gamma_2; \mathbf{B}).$$

Now (3.12.16) with $z_1 = -i, z_2 = 1, z_3 = i$ and the fact that $\overline{h}$ fixes the points $i, -1, -i$ imply that the distance between any pair of the points $h(1)$, $h(i)$, $h(-1)$, $h(-i)$ is at least

$$2d = \frac{2}{\sqrt{b^2+1}} \le \sqrt{2}.$$

Since $\text{length}(\gamma_j) \le \pi$, γ_j contains at most two points of $\{1, i, -1, -i\}$, $h(\gamma_j)$ contains at most two points of $\{h(1), h(i), h(-1), h(-i)\}$, and

$$\text{length}(h(\gamma_j)) \le 2(\pi - d)$$

for $j = 1, 2$. From this it follows that

$$1 \le \frac{\text{length}(h(\gamma_1))}{|h(z_1) - h(z_2)|}, \frac{\text{length}(h(\gamma_2))}{|h(z_2) - h(z_3)|} \le \frac{\pi - d}{\sin(\pi - d)} = m,$$

where $m = m(d)$. Thus

$$\frac{\omega(0, h(\gamma_1); \mathbf{B})}{\omega(0, h(\gamma_2); \mathbf{B})} = \frac{\text{length}(h(\gamma_1))}{\text{length}(h(\gamma_2))} \le mb$$

and we conclude that D is quasisymmetric with respect to harmonic measure.

Suppose next that D satisfies the harmonic quasisymmetry condition with respect to the points $w_0 \in D$ and $w_0^* \in D^*$. We introduce conformal mappings f and g as before and let h be the induced homeomorphism

$$h = g \circ f^{-1} : \partial\mathbf{B} \to \partial\mathbf{B}.$$

Then because D satisfies the harmonic quasisymmetry condition, it is easy to see that there exists a constant b such that (3.12.10) holds whenever $|z_1 - z_2| = |z_2 - z_3|$.

Next if $Q = D(w_1, w_2, w_3, w_4)$ is a quadrilateral in D with $\text{mod}(Q) = 1$, then $\text{mod}(f(Q)) = 1$ and $[z_1, z_2, z_3, z_4] = 2$ where $z_j = f(w_j)$, by Corollary 3.10.3. By Lemma 3.12.9, inequality (3.12.11) holds with $a = a(b)$. Hence by Lemma 3.10.4

$$\text{mod}(Q^*) = \text{mod}(\mathbf{B}(h(z_4), h(z_3), h(z_2), h(z_1))) \leq c$$

where $c = c(a)$ since $[h(z_1), h(z_2), h(z_3), h(z_4)] = [h(z_4), h(z_3), h(z_2), h(z_1)]$ is bounded. We conclude that D satisfies the conjugate quadrilateral inequality. $\square$

CHAPTER 4

Injectivity criteria

Suppose that f is a mapping which is locally injective in a domain D. When can we conclude that f is globally injective in D? The answer depends, of course, on the nature of the function f as well as on the domain D.

We consider here three classes of functions f as well as appropriate measures of growth which are necessary and which are sufficient for the functions in each class to be injective whenever D is a disk or half-plane. We then observe that the same kinds of conclusions hold if and only if D is a quasidisk.

We conclude with a characterization of the Jacobian of a conformal mapping f of D which holds whenever D is a quasidisk.

4.1. Meromorphic functions

We adopt the convention that a *meromorphic* function need not have poles and may be defined in a neighborhood of ∞ and that an *analytic* function is a meromorphic function without poles defined in a domain in $\mathbf{R}^2$. We consider here two criteria for the injectivity of a function f in these two classes involving

1° the Schwarzian derivative S_f of f,

2° the pre-Schwarzian derivative T_f of f.

We then show how each of these criteria can be used to characterize quasidisks.

The *Schwarzian* derivative of a function f meromorphic and locally injective in D is given by

$$S_f = \left(\frac{f''}{f'}\right)' - \frac{1}{2}\left(\frac{f''}{f'}\right)^2,$$

where we employ the usual convention regarding points in $\{\infty, f^{-1}(\infty)\}$.

If f is a Möbius transformation

$$f(z) = \frac{a\,z+b}{c\,z+d}, \qquad ad-bc \neq 0,$$

then

$$\frac{f''(z)}{f'(z)} = -\frac{2\,c}{c\,z+d}, \qquad \left(\frac{f''(z)}{f'(z)}\right)' = \frac{2\,c^2}{(c\,z+d)^2}$$

and hence $S_f(z) = 0$ for all z. Conversely if f is meromorphic with $S_f(z) = 0$ in a domain D, then f is the restriction of a Möbius transformation to D in which case f is injective in D.

If f is meromorphic in D and g is meromorphic in $f(D)$, then

$$S_{g\circ f}(z) = S_g(f(z))\, f'(z)^2 + S_f(z) \tag{4.1.1}$$

in D. In particular, whenever g is a Möbius transformation,

$$S_{g\circ f}(z) = S_f(z). \tag{4.1.2}$$

The following shows that the size of the Schwarzian relative to the hyperbolic metric is related to the global injectivity of a meromorphic function.

THEOREM 4.1.3 (Lehto [**114**], Nehari [**137**]). *If f is meromorphic and injective in a simply connected domain D, then*

$$\sup_{z\in D} |S_f(z)|\,\rho_D(z)^{-2} \le 3.$$

The constant 3 *is sharp.*

Thus the Schwarzian derivative S_f is bounded in absolute value by a multiple of the square of the hyperbolic density ρ_D whenever f is injective in a simply connected domain D. It is natural to ask if there is a sufficient condition, corresponding to this necessary condition, for injectivity.

This is indeed the case whenever D is a disk or half-plane.

THEOREM 4.1.4 (Nehari [**137**]). *If D is a disk or half-plane and if f is meromorphic with*

$$\sup_{z\in D} |S_f(z)|\,\rho_D(z)^{-2} \le 1/2,$$

then f is injective in D. The constant $1/2$ *is sharp.*

It is hence natural to ask for which domains the above result holds with $1/2$ replaced by some positive constant.

DEFINITION 4.1.5. We let $\sigma(D)$ denote the supremum of the constants $a \ge 0$ such that f is injective whenever f is meromorphic and locally injective in D with

$$\sup_{z\in D} |S_f(z)|\,\rho_D(z)^{-2} \le a. \tag{4.1.6}$$

Then $\sigma(D) \le 1/2$ for all domains D and D is a disk or half-plane if and only if $\sigma(D) = 1/2$ (Lehtinen [**112**]).

We observe next that *supremum* may be replaced by *maximum* in the definition of $\sigma(D)$.

THEOREM 4.1.7 (Lehto [**114**]). *If f is meromorphic in a simply connected domain D with*

$$\sup_{z\in D} |S_f(z)|\,\rho_D(z)^{-2} \le \sigma(D), \tag{4.1.8}$$

then f is injective in D.

SKETCH OF PROOF. Suppose that f is meromorphic in D with

$$\sup_{z\in D} |S_f(z)|\,\rho_D(z)^{-2} \le \sigma(D),$$

fix points $z_1, z_2, z_3 \in D$ so that $f(z_1), f(z_2), f(z_3)$ are distinct, and choose $r_j \in (0,1)$ so that $r_j \to 1$. Then for each j there exists a function f_j meromorphic in D such that

$$S_{f_j}(z) = r_j\, S_f(z)$$

in D; see, for example, Lehto [**116**]. Moreover by (4.1.2) we may assume that $f_j(z_k) = f(z_k)$ for $k = 1, 2, 3$. Then

$$\sup_{z\in D} |S_{f_j}(z)|\,\rho_D(z)^{-2} \le r_j\,\sigma(D) < \sigma(D),$$

the functions f_j are injective in D, and a subsequence converges locally uniformly to f. Hence f is also injective in D. □

The conclusion in Theorem 4.1.7 can be strengthened when D is a disk or half-plane. In this case f has a continuous extension to $\overline{D}$ and $f(D)$ is either a Jordan domain or the image of the strip domain

$$\{z = x + i\,y : |x| < \infty, |y| < 1\}$$

under a Möbius transformation (Gehring-Pommerenke [**69**]). It would be interesting to know what the situation is for other domains D.

The following result characterizes the domains D for which $\sigma(D) > 0$.

THEOREM 4.1.9 (Ahlfors [**5**], Gehring [**49**]). *A simply connected domain D is a K-quasidisk if and only if $\sigma(D) > 0$, where K and $\sigma(D)$ depend only on each other.*

Theorem 4.1.9 allows us to obtain a still stronger conclusion in Theorem 4.1.7 when (4.1.8) holds with strict inequality.

THEOREM 4.1.10. *If D is a quasidisk and if f is meromorphic in D with*

$$\sup_{z \in D} |S_f(z)|\, \rho_D(z)^{-2} \le c < \sigma(D),$$

then f has a K-quasiconformal extension to $\overline{\mathbf{R}}^2$ where K depends only on $\sigma(D)$ and c.

PROOF. Let $D' = f(D)$ and suppose that g is meromorphic in D' with

$$|S_g(w)|\, \rho_{D'}(w)^{-2} \le \sigma(D) - c.$$

Then

$$S_{g \circ f}(z) = S_g(f(z))\, f'(z)^2 + S_f(z), \qquad \rho_D(z) = \rho_{D'}(f(z))\, |f'(z)|,$$

whence

$$|S_{g \circ f}(z)|\, \rho_D(z)^{-2} \le |S_g(f(z))|\, \rho_{D'}(f(z))^{-2} + |S_f(z)|\, \rho_D(z)^{-2} \le \sigma(D).$$

Thus $g \circ f$ is injective in D, g is injective in D', and

$$\sigma(D') \ge \sigma(D) - c > 0.$$

Hence D and D' are K'-quasidisks, where K' depends only on $\sigma(D)$ and $\sigma(D) - c$, and f has a homeomorphic extension f^* which maps $\overline{D}$ onto $\overline{D'}$.

Next there exist K'-quasiconformal self-mappings ϕ and ψ of $\overline{\mathbf{R}}^2$ which map the upper half-plane $\mathbf{H}$ onto D and D', respectively. If r denotes reflection in the real axis, then $h_1 = \phi \circ r \circ \phi^{-1}$ and $h_2 = \psi \circ r \circ \psi^{-1}$ are quasiconformal reflections in ∂D and $\partial D'$ and $h_2 \circ f \circ h_1^{-1}$ defines a K-quasiconformal extension of f to $\overline{\mathbf{R}}^2$ where $K = (K')^4$. □

We consider next counterparts of the above results for the *pre-Schwarzian derivative*

$$T_f = \frac{f''}{f'}$$

of a function f analytic and locally injective in $D \subset \mathbf{R}^2$. In this case $T_f = 0$ in D if and only if f is a similarity mapping, in which case f is injective.

We then have the following analogue for the pre-Schwarzian derivative of Theorem 4.1.3.

THEOREM 4.1.11 (Osgood [**141**]). *If f is analytic and injective in a simply connected domain $D \subset \mathbf{R}^2$, then*

$$\sup_{z \in D} |T_f(z)| \, \rho_D(z)^{-1} \leq 4.$$

The constant 4 is sharp.

For which domains D can we reverse the implication in Theorem 4.1.11?

DEFINITION 4.1.12. We let $\tau(D)$ denote the supremum of the constants $b \geq 0$ such that f is injective whenever f is analytic and locally injective in $D \subset \mathbf{R}^2$ with

$$\sup_{z \in D} |T_f(z)| \, \rho_D(z)^{-1} \leq b. \tag{4.1.13}$$

Then $\tau(D) \leq 1/2$ for all domains D (Stowe [**155**]) and equality holds whenever D is a disk or half-plane (Becker-Pommerenke [**22**]). However in this case the converse does not hold; there exists a domain D with $\tau(D) = 1/2$ which is not a disk or half-plane (Stowe [**155**]).

On the other hand, we have the following analogue for Theorem 4.1.9.

THEOREM 4.1.14 (Astala-Gehring [**14**]). *A simply connected domain D is a K-quasidisk if and only if $\tau(D) > 0$, where K and $\tau(D)$ depend only on each other.*

Though the constants $\sigma(D)$ and $\tau(D)$ reflect the geometry of the domain D, little is known about their values except for $\sigma(D)$ in the following special cases.

1° If D is a sector of angle α, then

$$\sigma(D) = \frac{1}{2} \min \left(\frac{\alpha^2}{\pi^2}, \frac{2\pi\alpha - \alpha^2}{\pi^2} \right).$$

2° If D is a regular n-sided polygon, then

$$\sigma(D) = \left(\frac{n-2}{n} \right)^2.$$

3° Finally, if D is a rectangle with side ratio r, then

$$\sigma(D) = 1/8$$

for $.657 < r < 1.523$. In particular, $\sigma(D)$ does not always depend analytically on the shape of D.

See Lehto [**115**] for 1°, see Calvis [**30**], Lehtinen [**113**] for 2°, and see Miller-Van Wieren [**131**] for 3°.

4.2. Locally bilipschitz mappings

We consider next some analogues for bilipschitz functions of the injectivity and extension results established in the previous section.

DEFINITION 4.2.1. The mapping f is a *locally L-bilipschitz mapping* of $E \subset \mathbf{R}^2$ if each point of E has a neighborhood U such that

$$\frac{1}{L} |z_1 - z_2| \leq |f(z_1) - f(z_2)| \leq L \, |z_1 - z_2|$$

for $z_1, z_2 \in E \cap U$.

Suppose that f is a locally L-bilipschitz mapping of D. Then whether or not one can conclude that f is injective depends on L and on D.

EXAMPLE 4.2.2. If f is locally 1-bilipschitz in a domain $D \subset \mathbf{R}^2$, then f is injective in D. On the other hand, for each $L > 1$,

$$f(z) = \frac{|z|}{L} \exp(iL^2 \arg(z)), \qquad |\arg(z)| < \pi,$$

is a locally L-bilipschitz mapping of $D = \mathbf{R}^2 \setminus (-\infty, 0]$, which is not injective.

The following result shows that f is always injective for small L provided D is a disk or half-plane.

THEOREM 4.2.3 (John [**90**], [**91**]). *If D is a disk or half-plane and if f is a locally L-bilipschitz mapping of D with*

$$L \le 2^{1/4},$$

then f is injective in D.

PROOF. Suppose otherwise. Because f is a local homeomorphism, we can choose a disk U with $\overline{U} \subset D$ and points $z_1, z_2 \in \partial U$ such that f is injective in U with $f(z_1) = f(z_2)$.

Let β be the circular arc in $\overline{U}$ orthogonal to ∂U at z_1 and z_2 and let E denote the component of $U \setminus \beta$ for which $f(E)$ is enclosed by $f(\beta)$. Then

$$\text{length}(f(\beta)) \le L\, \text{length}(\beta)$$

because f is locally L-bilipschitz. Next

$$m(E) \le L^2\, m(f(E))$$

since f is injective and locally L-bilipschitz in U. Then from elementary geometry and the isoperimetric inequality we obtain

$$\frac{\text{length}(\beta)^2}{2\,\pi} < m(E) \le L^2\, m(f(E)) \le L^2\, \frac{\text{length}(f(\beta))^2}{4\,\pi} \le L^4\, \frac{\text{length}(\beta)^2}{4\,\pi},$$

whence $L^4 > 2$, a contradiction. □

The constant $2^{1/4}$ in Theorem 4.2.3 is not sharp. See, for example, Gevirtz [**71**] where it is shown that Theorem 4.2.3 holds with $(1+\sqrt{2})^{1/4}$ in place of $2^{1/4}$; $2^{1/2}$ is probably the right constant. The problem of determining this sharp bound has been open for more than forty years.

DEFINITION 4.2.4. We let $L(D)$ denote the supremum of the numbers $M \ge 1$ such that f is injective whenever f is locally L-bilipschitz in D with $L \le M$; D is *rigid* if $L(D) > 1$.

The following analogue of Theorem 4.1.9 describes the simply connected domains that are rigid.

THEOREM 4.2.5 (Gehring [**50**], Martio-Sarvas [**123**]). *A simply connected domain $D \subset \mathbf{R}^2$ is a K-quasidisk if and only if $L(D) > 1$, where K and $L(D)$ depend only on each other.*

The following counterpart of Theorem 4.1.10 for bilipschitz functions is a consequence of the above result.

THEOREM 4.2.6 (Gehring [**50**]). *If $D \subset \mathbf{R}^2$ is a quasidisk and if f is locally L-bilipschitz in D with $L < L(D)$, then f has an M-bilipschitz extension to $\overline{\mathbf{R}}^2$ where M depends only on L and $L(D)$.*

SKETCH OF PROOF. Let $D' = f(D)$ and suppose that g is locally L'-bilipschitz in D' with

$$L' < \frac{L(D)}{L}.$$

Then $g \circ f$ is locally LL'-bilipschitz in D with $LL' < L(D)$. Thus $g \circ f$ and f are injective in D, g is injective in D', and

$$L(D') \geq \frac{L(D)}{L} > 1.$$

Theorem 4.2.5 then implies that D and D' are K-quasidisks where K depends only on $L(D)$ and $L(D)/L$.

Next by Theorem 3.4.5 there exists a constant c, which depends only on K, such that $z_1, z_2 \in D$ and $w_1, w_2 \in D'$ can be joined by hyperbolic segments $\gamma \subset D$ and $\gamma' \subset D'$ where

$$\operatorname{length}(\gamma) \leq c\,|z_1 - z_2|, \qquad \operatorname{length}(\gamma') \leq c\,|w_1 - w_2|.$$

This and the fact that f is injective and locally L-bilipschitz in D imply that

$$|f(z_1) - f(z_2)| \leq \operatorname{length}(f(\gamma)) \leq L \operatorname{length}(\gamma) \leq Lc\,|z_1 - z_2|$$

and

$$|f^{-1}(w_1) - f^{-1}(w_2)| \leq \operatorname{length}(f^{-1}(\gamma')) \leq L \operatorname{length}(\gamma') \leq Lc\,|w_1 - w_2|.$$

Thus f is Lc-bilipschitz in D and hence has a homeomorphic extension $f^* : \overline{D} \to \overline{D'}$.

If D is unbounded, then so is D' and, as in the proof of Theorem 4.1.10, Theorem 2.1.8 implies that f^* has an M-bilipschitz extension to $\overline{\mathbf{R}}^2$ where M depends only on K and L. If D is bounded, then we can choose an auxiliary Möbius transformation ϕ so that $\phi(D)$ and $\phi(D')$ are unbounded and complete the proof as above. See Gehring [**50**] for the details. □

Theorem 4.2.6 has the following physical interpretation. Suppose that D is an elastic body in $\mathbf{R}^2$ and let f denote the deformation of D under a force field. Then

$$L_f(z) = \limsup_{h \to 0} \max\left(\frac{|f(z+h) - f(z)|}{|h|}, \frac{|h|}{|f(z+h) - f(z)|}\right)$$

measures the *strain* in D at the point z caused by the force field, f is locally bilipschitz if and only if $L_f(z)$ is bounded, and $L(D)$ is the supremum of allowable strains before D collapses. Theorem 4.2.6 says that if

$$\sup_{z \in D} L(z) < L(D),$$

then the shape of the deformed body $f(D)$ is roughly the same as that of the original body D.

The related problem in higher dimensions is considered in Martio-Sarvas [**123**].

4.3. Locally quasiconformal mappings

A mapping f is *locally* K-quasiconformal in a domain D if each point of D has a neighborhood in which f is K-quasiconformal. We consider here analogues of the preceding injectivity results for locally quasiconformal mappings. In order to do so we must first find a substitute for the quantities

$$\sup_{z \in D} |S_f(z)|\,\rho_D(z)^{-2}, \qquad \sup_{z \in D} |T_f(z)|\,\rho_D(z)^{-1}, \qquad \sup_{z_1, z_2 \in D} \frac{|f(z_1) - f(z_2)|}{|z_1 - z_2|}$$

for locally quasiconformal mappings. It turns out that the BMO (bounded mean oscillation) norm of the Jacobian J_f of a locally quasiconformal mapping f is a natural counterpart for each of the above quantities.

Suppose that u is a function locally integrable in a domain $D \subset \mathbf{R}^2$. Then the BMO-*norm* of u in D is given by

$$\|u\|_{\mathrm{BMO}(D)} = \sup_{B_0} \frac{1}{m(B_0)} \int_{B_0} |u - u_{B_0}|\, dm \tag{4.3.1}$$

where the supremum is taken over all disks B_0 with $\overline{B}_0 \subset D$ and

$$u_{B_0} = \frac{1}{m(B_0)} \int_{B_0} u\, dm.$$

The following observation suggests how this norm is related to the hyperbolic metric.

LEMMA 4.3.2. *If B_0 is a disk in D with center z_0, then*

$$\frac{1}{m(B_0)} \int_{B_0} h_D(z, z_0)\, dm \le 2.$$

PROOF. The left-hand side of the above inequality is invariant with respect to similarity mappings. Hence we may assume that B_0 is the unit disk $\mathbf{B}$. Then

$$\begin{aligned}\int_{\mathbf{B}} h_D(z,0)dm \le \int_{\mathbf{B}} h_{\mathbf{B}}(z,0)dm &= \int_{\mathbf{B}} \log\Big(\frac{1+|z|}{1-|z|}\Big) dm \\ &= \int_0^{2\pi} \Big(\int_0^1 \log\Big(\frac{1+t}{1-t}\Big)\, t\, dt \Big)\, d\theta \\ &= 2\pi \left[t + \frac{t^2-1}{2} \log\Big(\frac{1+t}{1-t}\Big)\right]_{t=0}^{t=1} = 2\, m(\mathbf{B}).\end{aligned}$$

□

With Lemma 4.3.2 we are able to derive the following useful estimate for the BMO norm of a harmonic function.

LEMMA 4.3.3. *If u is harmonic in $D \subset \mathbf{R}^2$, then*

$$\frac{1}{2}\, \|u\|_{\mathrm{BMO}(D)} \le \sup_{z \in D} |\mathrm{grad}\ u(z)|\, \rho_D^{-1}(z) \le 6 \|u\|_{\mathrm{BMO}(D)}$$

where ρ_D is the density of the hyperbolic metric h_D.

PROOF. Choose a disk $B_0 = \mathbf{B}(z_0, d)$ with $\overline{B}_0 \subset D$. Then since u is harmonic in D, $u(z_0) = u_{B_0}$ and it suffices to prove that

$$|\mathrm{grad}\ u(z_0)|\, \rho_D(z_0)^{-1} \le \frac{6}{m(B_0)} \int_{B_0} |u(z) - u(z_0)|\, dm \tag{4.3.4}$$

and

$$\frac{1}{m(B_0)} \int_{B_0} |u(z) - u(z_0)| dm \le 2 \sup_{z \in D} |\mathrm{grad}\ u(z)|\, \rho_D(z)^{-1}. \tag{4.3.5}$$

By performing a change of variable we may assume that $z_0 = 0$.

For (4.3.4) fix $0 < r < d$ and let

$$f(z) = \frac{1}{2\pi} \int_0^{2\pi} \frac{r\, e^{i\theta} + z}{re^{i\theta} - z}\, u(re^{i\theta})\, d\theta.$$

Then f is analytic in $\mathbf{B}(0,r)$ with $u = \mathrm{Re}(f)$ and differentiation yields

$$f'(0) = \frac{1}{\pi}\int_0^{2\pi} \frac{u(re^{i\theta})}{re^{i\theta}}\,d\theta = \frac{1}{\pi}\int_0^{2\pi} \frac{u(re^{i\theta}) - u(0)}{re^{i\theta}}\,d\theta.$$

Since $|\mathrm{grad}\, u(0)| = |f'(0)|$, we have

$$|\mathrm{grad}\, u(0)| \le \frac{1}{\pi}\int_0^{2\pi} r^{-1}|u(re^{i\theta}) - u(0)|\,d\theta.$$

From this together with inequality (3.2.1)

$$\begin{aligned}|\mathrm{grad}\, u(0)|\,\rho_D(0)^{-1} &\le |\mathrm{grad}\, u(0)|\,2d\\ &\le \frac{6}{m(B_0)}\int_0^d |\mathrm{grad}\, u(0)|\,\pi r^2 dr\\ &\le \frac{6}{m(B_0)}\int_{B_0} |u(z) - u(0)|\,dm.\end{aligned}$$

Next for (4.3.5) let γ be the hyperbolic geodesic joining 0 to z in D. Then

$$|u(z) - u(0)| \le c\int_\gamma \rho_D|dz| = c\,h_D(z,0),$$

where

$$c = \sup_{z\in D} |\mathrm{grad}\, u(z)|\,\rho_D(z)^{-1}$$

and

$$\frac{1}{m(B_0)}\int_{B_0} |u(z) - u(0)|dm \le \frac{c}{m(B_0)}\int_{B_0} h_D(z,0)\,dm \le 2c$$

by Lemma 4.3.2. □

COROLLARY 4.3.6 (Astala-Gehring [**14**]). *If f is analytic with $f' \neq 0$ in $D \subset \mathbf{R}^2$, then*

$$\frac{1}{4}\,\|\log J_f\|_{\mathrm{BMO}(D)} \le \sup_{z\in D} |T_f(z)|\,\rho_D(z)^{-1} \le 3\,\|\log J_f\|_{\mathrm{BMO}(D)}.$$

PROOF. Let $u(z) = \log|f'(z)|$. Then u is harmonic in D,

$$\log J_f(z) = 2\,\log|f'(z)| = 2\,u(z), \qquad |T_f(z)| = \frac{|f''(z)|}{|f'(z)|} = |\mathrm{grad}\, u(z)|,$$

and the desired conclusion follows from Lemma 4.3.3. □

Corollary 4.3.6 shows that the BMO-norm of $\log J_f$ is a natural alternative for the pre-Schwarzian derivative T_f when considering injectivity results for locally conformal mappings. Moreover the following quasiconformal counterpart of Theorem 4.1.3 and Theorem 4.1.11 suggests that this norm offers a way to extend results on the injectivity of analytic functions to the class of locally quasiconformal mappings.

THEOREM 4.3.7 (Reimann [**146**]). *If f is K-quasiconformal in D with $f(D) \subset \mathbf{R}^2$, then*

$$\|\log J_f\|_{\mathrm{BMO}(D)} \le m$$

where $m = m(K) < \infty$. Moreover, when $D = \mathbf{R}^2$,

$$\lim_{K\to 1} m(K) = 0.$$

The following quasiconformal analogue of Theorems 4.1.4 and 4.2.3 shows that the implication in Theorem 4.3.7 can also be reversed for locally K-quasiconformal mappings in disks and half-planes provided K is not too large.

THEOREM 4.3.8 (Astala-Gehring [**14**]). *Suppose that D is a disk or half-plane. Then for each $1 \leq K < 2$ there exists a constant $c = c(K) > 0$ with the following property. If f is locally K-quasiconformal in D with $f(D) \subset \mathbf{R}^2$ and*

$$\| \log J_f \|_{\mathrm{BMO}(D)} \leq c,$$

then f is injective in D. The constant 2 is sharp; i.e., no such constant c exists if $K \geq 2$.

Finally we see that the class of domains D for which such a converse exists coincides with the family of quasidisks.

THEOREM 4.3.9 (Astala-Gehring [**14**]). *A simply connected domain $D \subset \mathbf{R}^2$ is a quasidisk if and only if for some $K > 1$ there exists a constant $c > 0$ such that f is injective whenever f is locally K-quasiconformal in D with $f(D) \subset \mathbf{R}^2$ and*

$$\| \log J_f \|_{\mathrm{BMO}(D)} \leq c.$$

4.4. Jacobian of a conformal mapping

If f is conformal in $D \subset \mathbf{R}^2$ with $f(D) \subset \mathbf{R}^2$, then the function

$$u = \log J_f$$

is harmonic in D. Next Theorem 4.3.7 implies that u also has finite BMO-norm in D. Hence it is natural to ask under what circumstances these two properties characterize the Jacobian of a conformal mapping.

The answer yields still another description of a quasidisk.

THEOREM 4.4.1 (Astala-Gehring [**14**]). *A simply connected domain $D \subset \mathbf{R}^2$ is a quasidisk if and only if there exists a constant $c > 0$ such that each function u harmonic in D with*

$$\|u\|_{\mathrm{BMO}(D)} \leq c$$

can be written in the form

$$u = \log J_f \tag{4.4.2}$$

where f is conformal in D with $f(D) \subset \mathbf{R}^2$.

PROOF. Necessity follows from Theorem 4.3.9 above. For sufficiency suppose there exists a constant $c > 0$ such that (4.4.2) holds whenever u is harmonic with $\|u\|_{\mathrm{BMO}(D)} \leq c$. Next choose g analytic in D with $g'(z) \neq 0$ and

$$\left| \frac{g''(z)}{g'(z)} \right| \rho_D(z)^{-1} \leq \frac{c}{4}$$

in D and let $u = \log J_g$. Then (4.4.2) holds where f is conformal in D and

$$h(z) = \frac{g'(z)}{f'(z)}$$

is analytic with

$$2 \log |h(z)| = J_g(z) - J_f(z) = 0$$

in D. Hence h is constant in D, $g = af + b$ where a and b are constants, and g is injective in D. Thus

$$\tau(D) \geq \frac{c}{4} > 0$$

and D is a quasidisk by Theorem 4.1.14. □

It would be interesting to know if there exists an analogue of Theorem 4.4.1 for the Jacobian of a quasiconformal mapping.

CHAPTER 5

Criteria for extension

Suppose $\mathcal{F}$ denotes a certain property of functions, for example, continuity or integrability, and let $\mathcal{F}(D)$ be the family of functions defined on D with the property $\mathcal{F}$. When does each function in $\mathcal{F}(D)$ have an extension to $\mathbf{R}^2$ which is in $\mathcal{F}(\mathbf{R}^2)$? As in Chapter 4, the answer depends on the nature of $\mathcal{F}$ as well as the domain D.

We consider here four different properties $\mathcal{F}$ for which such an extension is possible whenever D is a disk or half-plane and then observe that the same is true if and only if D is a quasidisk. Two of these conditions deal with real-valued functions and the rest with the quasiconformal and bilipschitz classes.

5.1. Functions of bounded mean oscillation

We have already introduced the BMO-norm of the Jacobian J_f of a locally quasiconformal mapping f in order to study when f is injective. We consider now the family of functions for which this norm is finite.

Suppose that u is a locally integrable real-valued function in a domain $D \subset \mathbf{R}^2$. We say that u has *bounded mean oscillation* in D, or is in $\mathrm{BMO}(D)$ if

$$\|u\|_{\mathrm{BMO}(D)} = \sup_{B_0} \frac{1}{m(B_0)} \int_{B_0} |u - u_{B_0}| dm < \infty,$$

where as in (4.3.1), the supremum is taken over all disks B_0 with $\overline{B}_0 \subset D$ and

$$u_{B_0} = \frac{1}{m(B_0)} \int_{B_0} u \, dm.$$

Functions of bounded mean oscillation occur naturally in many parts of mathematics. They were first studied in connection with problems in elasticity (John, Nirenberg [**90**], [**93**]) and partial differential equations (Moser [**134**]). They were later found to play an important role in harmonic analysis. See Fefferman [**39**], Fefferman-Stein [**40**], Garnett [**44**], and Reimann-Rychener [**147**].

We have that

$$L^\infty(D) \subset \mathrm{BMO}(D). \tag{5.1.1}$$

It is easy to see that the inclusion is strict when D is simply connected.

LEMMA 5.1.2. *If $D \subset \mathbf{R}^2$ is simply connected, then for each $z_1 \in D$ the hyperbolic distance $u(z) = h_D(z, z_1)$ is in* $\mathrm{BMO}(D)$ *with*

$$\|u\|_{\mathrm{BMO}(D)} \le 4.$$

Hence, in particular, $u \in \mathrm{BMO}(D) \setminus L^\infty(D)$.

PROOF. If B_0 is any disk with center z_0 and $\overline{B}_0 \subset D$, then

$$
\begin{aligned}
|u_{B_0} - u(z_0)| &\le \frac{1}{m(B_0)} \int_{B_0} |h_D(z,z_1) - h_D(z_0,z_1)|\, dm \\
&\le \frac{1}{m(B_0)} \int_{B_0} h_D(z,z_0)\, dm \le 2
\end{aligned}
\tag{5.1.3}
$$

by Lemma 4.3.2, whence

$$\frac{1}{m(B_0)} \int_{B_0} |u(z) - u_{B_0}|\, dm \le \frac{1}{m(B_0)} \int_{B_0} (|u(z) - u(z_0)| + 2)\, dm \le 4.$$

□

If $v \in \mathrm{BMO}(\mathbf{R}^2)$ and if u is the restriction of v to D, then u is in $\mathrm{BMO}(D)$ with

$$\|u\|_{\mathrm{BMO}(D)} \le \|v\|_{\mathrm{BMO}(\mathbf{R}^2)}.$$

The following example shows that the converse is not always true for $\mathrm{BMO}(D)$ even though this is trivially true for the class $L^\infty(D)$ in (5.1.1).

EXAMPLE 5.1.4. If $u(z) = h_D(z,1)$ where D is the half-strip

$$D = \{z = x + iy : 0 < x < \infty,\ |y| < 1\},$$

then u is in $\mathrm{BMO}(D)$ but u has no extension v in $\mathrm{BMO}(\mathbf{R}^2)$.

Our proof of this depends on the following inequality, which will also be needed in Section 9.2.

LEMMA 5.1.5. *If u is in $\mathrm{BMO}(D)$ and if B_1, B_0 are disks with $B_1 \subset B_0 \subset D$, then*

$$|u_{B_1} - u_{B_0}| \le \frac{e}{2} \left(\log \frac{m(B_0)}{m(B_1)} + 1 \right) \|u\|_{\mathrm{BMO}(D)}.$$

PROOF. Suppose first that $m(B_0) \le e\, m(B_1)$. Since

$$\int_{B_0 \setminus B_1} (u - u_{B_0})\, dm + \int_{B_1} (u - u_{B_0})\, dm = \int_{B_0} (u - u_{B_0})\, dm = 0,$$

we obtain

$$|u_{B_0} - u_{B_1}| = \left| \frac{1}{m(B_1)} \int_{B_1} (u - u_{B_0})\, dm \right| = \left| \frac{1}{m(B_1)} \int_{B_0 \setminus B_1} (u - u_{B_0})\, dm \right|$$

and

$$|u_{B_0} - u_{B_1}| \le \frac{1}{2} \frac{1}{m(B_1)} \int_{B_0} |u - u_{B_0}|\, dm \le \frac{e}{2} \|u\|_{\mathrm{BMO}(D)}$$

by our assumption.

Next let k be the smallest integer for which $m(B_0) \le e^k\, m(B_1)$ and choose disks B_j so that

$$B_1 \subset B_2 \subset \cdots \subset B_{k+1} = B_0 \qquad \text{and} \qquad m(B_{j+1}) \le e\, m(B_j)$$

for $j = 1, 2, \ldots, k$. Then

$$
\begin{aligned}
|u_{B_0} - u_{B_1}| &\le \sum_{j=1}^{k} |u_{B_j} - u_{B_{j+1}}| \le \frac{k\, e}{2} \|u\|_{\mathrm{BMO}(D)} \\
&\le \frac{e}{2} \left(\log \frac{m(B_0)}{m(B_1)} + 1 \right) \|u\|_{\mathrm{BMO}(D)}
\end{aligned}
$$

by what was proved above and the choice of k. □

We turn now to a proof of the result in Example 5.1.4. Suppose that u has an extension v in $\mathrm{BMO}(\mathbf{R}^2)$, choose $n > 5$ so that

$$\|v\|_{\mathrm{BMO}(\mathbf{R}^2)} < \frac{1}{e}\,\frac{n-5}{2\log n + 1},$$

and let $B_j = \mathbf{B}(z_j, 1)$ and $B_0 = \mathbf{B}(n, n)$ where $z_1 = 1$ and $z_2 = 2n-1$. Then

$$|v_{B_j} - v_{B_0}| \le \frac{e}{2}\left(\log\frac{m(B_0)}{m(B_j)} + 1\right)\|v\|_{\mathrm{BMO}(\mathbf{R}^2)}$$

for $j = 1, 2$ by Lemma 5.1.5, whence

$$|u_{B_1} - u_{B_2}| = |v_{B_1} - v_{B_2}| < n - 5. \tag{5.1.6}$$

Next if γ is a hyperbolic segment joining z_1 and z_2 in D, then

$$|u(z_1) - u(z_2)| = h_D(z_1, z_2) = \int_\gamma \rho_D(z)\,ds \ge \frac{\mathrm{length}(\gamma)}{2} \ge n - 1$$

by inequality (3.2.1) and

$$|u_{B_1} - u_{B_2}| \ge |u(z_1) - u(z_2)| - |u(z_1) - u_{B_1}| - |u(z_2) - u_{B_2}| \ge n - 5$$

by (5.1.3), which contradicts (5.1.6).

The domain D in Example 5.1.4 is not a quasidisk by Example 1.4.5. On the other hand, functions in $\mathrm{BMO}(D)$ have extensions to $\mathrm{BMO}(\mathbf{R}^2)$ whenever D is a disk or half-plane.

THEOREM 5.1.7 (Reimann-Rychener [**147**]). *If D is a disk or half-plane and if $u \in \mathrm{BMO}(D)$, then u has an extension $v \in \mathrm{BMO}(\mathbf{R}^2)$ with*

$$\|v\|_{\mathrm{BMO}(\mathbf{R}^2)} \le c\,\|u\|_{\mathrm{BMO}(D)}$$

where $c \ge 1$ is an absolute constant.

DEFINITION 5.1.8. A domain D is a BMO-*extension domain* if there exists a constant $c \ge 1$ such that each $u \in \mathrm{BMO}(D)$ has an extension $v \in \mathrm{BMO}(\mathbf{R}^2)$ with

$$\|v\|_{\mathrm{BMO}(\mathbf{R}^2)} \le c\,\|u\|_{\mathrm{BMO}(D)}.$$

This extension property then yields another characterization for the class of quasidisks.

THEOREM 5.1.9 (Jones [**94**]). *A simply connected domain $D \subset \mathbf{R}^2$ is a K-quasidisk if and only if it is a* BMO*-extension domain with constant c, where K and c depend only on each other.*

5.2. Sobolev and finite energy functions

We assume throughout this section that u is locally integrable and ACL (absolutely continuous on lines) in a domain $D \subset \mathbf{R}^2$; see Definition 1.1.7. We say that u has *finite Dirichlet energy* or is in $L^2_1(D)$ if

$$E_D(u) = \int_D |\mathrm{grad}\, u|^2\,dm < \infty.$$

When do such functions u have an extension v in $L^2_1(\mathbf{R}^2)$?

EXAMPLE 5.2.1. If D is a disk or half-plane and if $u \in L_1^2(D)$, then u has an extension $v \in L_1^2(\mathbf{R}^2)$ with

$$E_{\mathbf{R}^2}(v) \le 2\, E_D(u).$$

However this result does not hold for all domains D. For example let D again be the half-strip

$$D = \{z = x + i\,y : 0 < x < \infty, |y| < 1\}$$

and for $j > 1$ and $z = x + i\,y \in D$ let

$$u_j(z) = \begin{cases} -1 & \text{if } 0 < x \le j-1, \\ x - j & \text{if } j-1 < x \le j+1, \\ 1 & \text{if } j+1 < x < \infty. \end{cases}$$

Then $E_D(u_j) = 4$.

Suppose that v_j is an extension of u_j in $L_1^2(\mathbf{R}^2)$ and let C_r denote the boundary of the square with corners at $(j-r, \pm r)$ and $(j+r, \pm r)$ for $1 < r < j$. Then v_j is absolutely continuous and assumes the values -1 and 1 on C_r for almost all $r \in (1, j)$. Thus

$$16 \le \left(\int_{C_r} |\text{grad } v_j(z)||dz| \right)^2 \le 8r \int_{C_r} |\text{grad } v_j(z)|^2 |dz|$$

for almost all $r \in (1, j)$ and

$$E_{\mathbf{R}^2}(v_j) \ge \int_1^j \left(\int_{C_r} |\text{grad } v_j(z)|^2 |dz| \right) dr \ge c_j\, E_D(u_j)$$

where

$$c_j = \frac{1}{2} \log j \to \infty$$

as $j \to \infty$. Hence there exists no constant c such that each $u \in L_1^2(D)$ has an extension $v \in L_1^2(\mathbf{R}^2)$ with

$$E_{\mathbf{R}^2}(v) \le c\, E_D(u).$$

DEFINITION 5.2.2. A domain D is an *L_1^2-extension domain* if there exists a constant $c \ge 1$ such that each function $u \in L_1^2(D)$ has an extension $v \in L_1^2(\mathbf{R}^2)$ with

$$E_{\mathbf{R}^2}(v) \le c\, E_D(u).$$

We see then that the simply connected domains with this property are quasidisks.

THEOREM 5.2.3 (Gol′dstein, Latfullin, Vodop′janov [**75**], [**76**]). *A simply connected domain D is a quasidisk if and only if it is an L_1^2-extension domain.*

The function u is in the Sobolev space $W_1^p(D)$, $1 \le p < \infty$, if

$$\|u\|_{W_1^p(D)} = \left(\int_D |u|^p \, dm \right)^{1/p} + \left(\int_D |\text{grad } u|^p \, dm \right)^{1/p} < \infty.$$

It is natural to ask if there is an analogue of Theorem 5.2.3 for the Sobolev class. This is, in fact, true for simply connected domains which are bounded.

DEFINITION 5.2.4. A domain D is a *Sobolev extension domain* if for each p, $1 \le p < \infty$, there exists a constant $c_p \ge 1$ such that each function u in $W_1^p(D)$ has an extension v in $W_1^p(\mathbf{R}^2)$ such that

$$\|v\|_{W_1^p(\mathbf{R}^2)} \le c_p \, \|u\|_{W_1^p(D)}.$$

THEOREM 5.2.5 (Jones [**95**]). *A bounded simply connected domain D is a quasidisk if and only if it is a Sobolev extension domain.*

5.3. Quasiconformal mappings

If D is a disk or half-plane, then each K-quasiconformal self-mapping f of D can be extended by reflection to yield a K-quasiconformal self-mapping g of $\overline{\mathbf{R}}^2$ (Lehto-Virtanen [**117**]).

DEFINITION 5.3.1. D is a *quasiconformal extension domain* if there exists a constant $c \ge 1$ such that each K-quasiconformal self-mapping f of D has a cK-quasiconformal extension g to $\overline{\mathbf{R}}^2$.

EXAMPLE 5.3.2. D is a disk or half-plane if and only if it is a quasiconformal extension domain with $c = 1$.

Our proof of this result depends on the following characterization for Jordan domains (Erkama [**36**], Hag [**77**]).

LEMMA 5.3.3. *A simply connected domain D is a Jordan domain if and only if every conformal self-mapping of D has a homeomorphic extension to $\overline{D}$.*

PROOF. If D is a Jordan domain, then every conformal self-mapping of D has a a homeomorphic extension to $\overline{D}$ by a well-known theorem of Carathéodory (Pommerenke [**145**]). For the converse we show that ∂D is locally connected with no cut points, and hence a Jordan curve (Pommerenke [**145**]), whenever each conformal self-mapping of D has a homeomorphic extension to $\overline{D}$.

For the local connectivity, suppose that g is a conformal map of the unit disk $\mathbf{B}$ onto D. Then there exists a point $z_0 \in \partial\mathbf{B}$ at which g has a radial limit (Pommerenke [**145**]). Let

$$\lim_{r \to 1} g(r\, z_0) = w_0.$$

Next for each $z_1 \in \partial\mathbf{B}$ let $h(z) = \dfrac{z_1}{z_0}\, z$. Then by hypothesis, the mapping

$$f = g \circ h \circ g^{-1} : D \to D$$

has a homeomorphic extension $f^* : \overline{D} \to \overline{D}$. In particular,

$$\lim_{r \to 1} g(r\, z_1) = \lim_{r \to 1} g \circ h(r\, z_0) = \lim_{r \to 1} f^* \circ g(r\, z_0) = f^*(w_0).$$

Since z_1 is an arbitrary point in $\partial\mathbf{B}$, we conclude that g has a radial limit at each point of $\partial\mathbf{B}$. Arguing again as above, we see that g has a limit at each point of $\partial\mathbf{B}$. Thus g has a continuous extension to $\overline{\mathbf{B}}$ and ∂D is locally connected (Pommerenke [**145**]). Finally, there exists $w_0 \in \partial D$ which is not a cut point of ∂D (Pommerenke [**145**]). But this implies that ∂D is cut point free. For let w_1 be any point in $\partial D \setminus \{w_0\}$ and choose $z_0, z_1 \in \partial\mathbf{B}$ such that $g(z_0) = w_0$ and $g(z_1) = w_1$. As before, $w_1 = f^*(w_0)$,

$$\partial D \setminus \{w_1\} = f^*(\partial D) \setminus \{f^*(w_0)\} = f^*(\partial D \setminus \{w_0\})$$

is connected, and hence w_1 is also not a cut point of ∂D. □

We turn now to the proof of the result in Example 5.3.2. For sufficiency we may assume that D is bounded since D is a Jordan domain by the lemma above. Next choose a disk $B \subset D$ with two boundary points $z_1, z_2 \in \partial D$ and choose $w_1, w_2 \in \partial D$ such that

$$|w_1 - w_2| = \operatorname{diam}(D).$$

Because D is a Jordan domain, there exists a conformal self-mapping f of D such that $f^*(z_j) = w_j$ for $j = 1, 2$ where f^* is the homeomorphic extension of f to $\overline{D}$. By hypothesis, f^* is the restriction of a Möbius transformation to $\overline{D}$ and hence $f(B)$ is a disk in D with

$$\operatorname{diam}(f(B)) = \operatorname{diam}(D).$$

Hence $f(B) = D$.

The following quasiconformal counterpart of Example 5.3.2 yields another characterization for quasidisks.

THEOREM 5.3.4 (Gehring-Hag [**55**], Rickman [**149**]). *A simply connected domain D is a K-quasidisk if and only if it is a quasiconformal extension domain with constant c, where K and c depend only on each other.*

5.4. Bilipschitz mappings

Suppose that D is a disk, that $C = \partial D$, and that $f : C \to C'$ is a homeomorphism. Then C' is the boundary of a Jordan domain D'. A well-known theorem of Schoenflies (Newman [**140**]) asserts that f has a homeomorphic extension g to D which maps D onto D'. What is the analogue of this result for bilipschitz mappings?

DEFINITION 5.4.1. A Jordan domain D with $C = \partial D$ has the *bilipschitz extension property* if there exists a constant $c \geq 1$ such that each L-bilipschitz mapping $f : C \to C'$ has a cL-bilipschitz extension $g : D \to D'$.

EXAMPLE 5.4.2. For each $L > 1$ there exists a bounded Jordan domain D with $C = \partial D$ and an L-bilipschitz mapping f of C which has no M-bilipschitz extension to D for any constant $M < \infty$.

To see this, fix $L > 1$ and let

$$D = \{z = x + iy : |x| < 1,\ a(|x|^{1/2} - 1) < y < 1\}$$

where $a = (L-1)/2L$. Then D has an outward directed cusp with tip at $z = -ia$ and

$$f(x + iy) = x + i|y|$$

is L-bilipschitz on $C = \partial D$. Next if g is an M-bilipschitz extension of f to D and if

$$\gamma = \gamma_t = \{z = x + iy \in D :\ y = t\}$$

for $-a < t < 0$, then $g(\gamma) \subset D$ and

$$M \geq \frac{\operatorname{length}(g(\gamma))}{\operatorname{length}(\gamma)} \geq \frac{a^2}{a+t} \to \infty$$

as $t \to -a$, a contradiction.

THEOREM 5.4.3 (Gehring [**50**], Tukia [**160**], [**162**]). *A bounded Jordan domain D is a K-quasidisk if and only if it has the bilipschitz extension property with constant c, where K and c depend only on each other.*

Necessity also holds in the above theorem when D is unbounded. However sufficiency fails in this case. For example, the half-strip

$$D = \{z = x + i\,y : \; 0 < x < \infty, \; |y| < 1\}$$

has the bilipschitz extension property but D is not a quasidisk by Example 1.4.5.

CHAPTER 6

Two-sided criteria

Suppose that a property of a domain D is not enough to guarantee that it is a quasidisk. Could it still be true that if D and its exterior domain D^* both have this property, then D, and hence D^*, is a quasidisk? We consider here five different properties for which this question has an affirmative answer.

6.1. Linear local connectivity revisited

We recall the properties 1° and 2° in Definition 2.4.1. For a set E there is to exist a constant $c \geq 1$ such that

1° $E \cap \mathbf{B}(z_0, r)$ lies in a component of $E \cap \mathbf{B}(z_0, cr)$ and
2° $E \setminus \overline{\mathbf{B}}(z_0, r)$ lies in a component of $E \setminus \overline{\mathbf{B}}(z_0, r/c)$

for each $z_0 \in \mathbf{R}^2$ and each $r > 0$.

These properties are in a sense dual and play similar roles. Domains with one of these properties appear quite often, and the terms *linearly connected domains* and *John disks* have been introduced for simply connected domains with nondegenerate boundary possessing property 1° and 2°, respectively. Linearly connected domains and John disks have been studied extensively by Pommerenke [**145**], Näkki and Väisälä [**136**], and others.

The following results show that these domains may be viewed as one-sided quasidisks.

THEOREM 6.1.1 (Gehring-Martio [**64**]). *A simply connected domain D is a quasidisk if and only if D and D^* both have property* 1° *in Definition* 2.4.1.

PROOF. If D is a quasidisk, then so is D^*, and it follows from Theorem 2.4.4 that both D and D^* have property 1°.

To prove the converse, we note first that D^* is simply connected as the complement of a connected set. Next ∂D (or ∂D^*) is a Jordan curve by the converse of Jordan's curve theorem if $\infty \in D^*$ (or D). If $\infty \in \partial D$, then ∂D must be locally connected and free of cut points in $\mathbf{R}^2$. Since D^* is a domain, ∞ is not a cut point. Hence ∂D is a Jordan curve. For more details see Newman [**140**] and Pommerenke [**145**].

See Figure 6.1. Assume that $z_1, z_2 \in D \setminus \overline{\mathbf{B}}(z_0, r)$ cannot be joined in $D \setminus \overline{\mathbf{B}}(z_0, r/c)$, and let $r' > r$ so that $z_1, z_2 \in D \setminus \overline{\mathbf{B}}(z_0, r')$. Then z_1, z_2 are separated in D by $\partial \mathbf{B}(z_0, r/c)$ and hence by an open arc α of $\partial \mathbf{B}(z_0, r/c)$. See Theorem VI.7.1 in Newman [**140**]. Let w_1 and w_2 in $\partial D = \partial D^*$ denote the endpoints of $\overline{\alpha}$. Now pick w_1', w_2' in $D^* \cap \mathbf{B}(z_0, r'/c)$ so that w_1, w_2 are accessible from w_1', w_2' in $D^* \cap \mathbf{B}(z_0, r'/c)$. These points can in turn be joined in $D^* \cap \mathbf{B}(z_0, r')$ and we have an open arc β in $D^* \cap \mathbf{B}(z_0, r')$ joining w_1 and w_2. The Jordan curve $\overline{\alpha} \cup \beta$ in $\mathbf{B}(z_0, r')$ must separate z_1 and z_2 in contradiction to the fact that z_1, z_2 both lie outside $\mathbf{B}(z_0, r')$. □

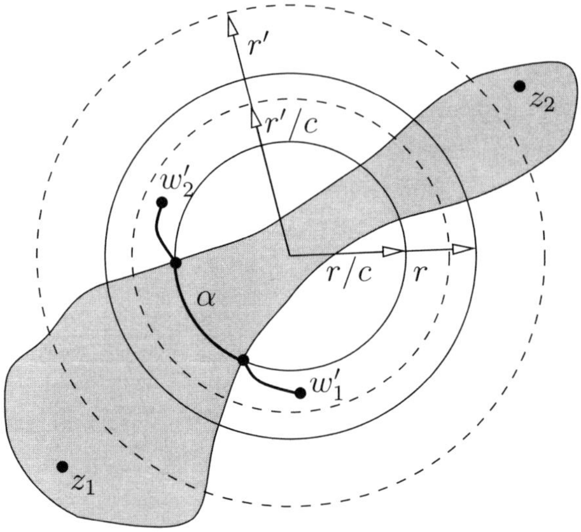

FIGURE 6.1

The corresponding result for property 2° does not follow for merely simply connected domains since a John disk need not be a Jordan domain. A disk minus a radius provides a standard counterexample.

THEOREM 6.1.2 (Näkki-Väisälä [**136**]). *A Jordan domain D is a quasidisk if and only if D and D^* are John disks.*

6.2. Hardy-Littlewood property

Suppose that D is a domain in $\mathbf{R}^2$ and that f is a real- or complex-valued function defined in D. We say that f is in the *Lipschitz class* $\text{Lip}_\alpha(D)$, $0 < \alpha \le 1$, if there exists a constant m such that

$$|f(z_1) - f(z_2)| \le m|z_1 - z_2|^\alpha \tag{6.2.1}$$

for all z_1 and z_2 in D, and we let $\|f\|_\alpha$ denote the infimum of the numbers m so that (6.2.1) holds. If there exists a constant m such that (6.2.1) holds in all disks $B \subset D$, we say that f lies in the *local Lipschitz class* $\text{loc}\,\text{Lip}_\alpha(D)$, and $\|f\|_\alpha^{\text{loc}}$ denotes the corresponding infimum over all such m.

For an analytic function satisfying (6.2.1) it follows from the Cauchy integral formula for the derivative that

$$|f'(z)| \le m\,\text{dist}(z, \partial D)^{\alpha-1}. \tag{6.2.2}$$

Conversely, the following well-known result of Hardy and Littlewood shows that condition (6.2.2) is also sufficient to guarantee that f lies in $\text{Lip}_\alpha(D)$ when D is a disk. See, for example, Theorem 5.1 in [**34**].

THEOREM 6.2.3. *If f is analytic in a disk $B = \mathbf{B}(z_0, r)$ with*

$$|f'(z)| \le m\,\text{dist}(z, \partial B)^{\alpha-1},$$

then $f \in \mathrm{Lip}_\alpha(B)$ *with*

$$\|f\|_\alpha \le \frac{c_0 m}{\alpha}$$

where c_0 *is an absolute constant.*

This result can be extended to uniform domains ([**64**]) and hence to quasidisks by Theorem 3.4.5.

THEOREM 6.2.4 (Gehring-Martio [**64**]). *Suppose* D *is a uniform domain in* $\mathbf{R}^2$. *If* f *is an analytic function in* D *which satisfies*

$$|f'(z)| \le m \operatorname{dist}(z, \partial D)^{\alpha-1} \tag{6.2.5}$$

for $z \in D$ *and* $\alpha \in (0,1]$, *then*

$$|f(z_1) - f(z_2)| \le \frac{cm}{\alpha}|z_1 - z_2|^\alpha$$

for $z_1, z_2 \in D$. *Here* c *is a constant that depends only on the domain* D.

PROOF. Fix $z_1, z_2 \in D$. Since D is uniform, there is a curve γ joining z_1 and z_2 in D satisfying

$$\operatorname{length}(\gamma) \le a|z_1 - z_2|,$$
$$\min_{j=1,2} \operatorname{length}(\gamma_j) \le a \operatorname{dist}(z, D), \ z \in \gamma,$$

where γ_1, γ_2 are the components of $\gamma \setminus \{z\}$. Then

$$\begin{aligned} |f(z_1) - f(z_2)| &\le \int_\gamma |f'(\zeta)|\,|d\zeta| \\ &\le m \int_\gamma \operatorname{dist}(\zeta, \partial D)^{\alpha-1}\,|d\zeta| \\ &\le 2ma^{1-\alpha} \int_0^{\operatorname{length}(\gamma)/2} s^{1-\alpha} ds = \frac{2ma^{1-\alpha}}{\alpha}\left(\frac{\operatorname{length}(\gamma)}{2}\right)^\alpha \\ &\le 2^{1-\alpha} a \frac{m}{\alpha}|z_1 - z_2|^\alpha \le \frac{cm}{\alpha}|z_1 - z_2|^\alpha \end{aligned}$$

where $c = 2a$. □

DEFINITION 6.2.6. We say that a proper subdomain D of $\mathbf{R}^2$ has the *Hardy-Littlewood property* if there exists a constant $c = c(D)$ such that for $0 < \alpha \le 1$, f is in $\mathrm{Lip}_\alpha(D)$ with $\|f\|_\alpha \le c/\alpha$ whenever f is analytic with

$$|f'(z)| \le \operatorname{dist}(z, \partial D)^{\alpha-1} \tag{6.2.7}$$

in D. Following Kaufman-Wu [**99**] we say that D has the *Hardy-Littlewood property of order* α if there is a constant $k = k(D, \alpha)$ such that if f satisfies (6.2.7), then $\|f\|_\alpha \le k$.

REMARK 6.2.8. If D has the Hardy-Littlewood property, then it has the Hardy-Littlewood property of any order α, while the converse is not true. See [**15**] for a counterexample.

By Theorem 6.2.4 all uniform domains and hence all quasidisks have the Hardy-Littlewood property. Unfortunately this property does not characterize quasidisks directly, as the following example due to Lappalainen [**111**] shows. A similar example is constructed by Gehring and Martio in [**66**, 2.26].

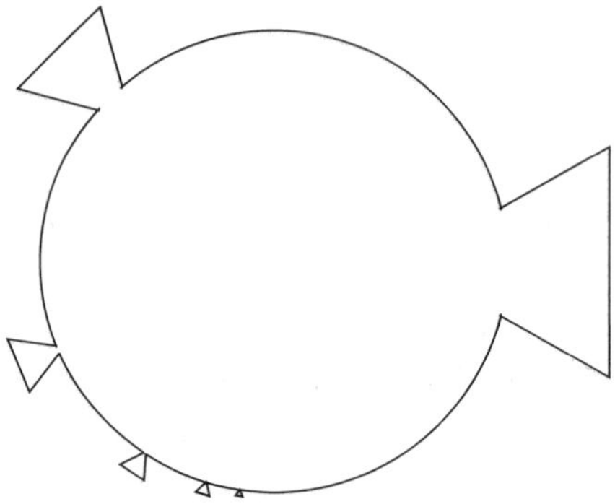

FIGURE 6.2

EXAMPLE 6.2.9. Define a Jordan domain D as

$$D = \mathbf{B} \cup \bigcup_{i=0}^{\infty} \Delta_i,$$

where Δ_i is the interior of an equilateral triangle with side length 2^{-i}. The polar angle θ_i for the center and the vertex of Δ_i closest to the origin is $3\pi(1 - 2^{-i})/2$ and the distance from this vertex to the origin is $r_i = 1 - 4^{-i}/2$. Then D has the Hardy-Littlewood property but is not a quasidisk since it violates the three-point condition in Section 2.2.

To see that D has the Hardy-Littlewood property, let f be an analytic function satisfying (6.2.5) in D and choose any z_1, z_2 in D. Now $\mathbf{B}$ and each Δ_i are quasidisks and hence uniform domains by Theorem 3.4.5. Thus if z_1 and z_2 both lie in $\mathbf{B}$ or in some Δ_i, Theorem 6.2.4 shows that f lies in $\mathrm{Lip}_\alpha(\mathbf{B})$ or $\mathrm{Lip}_\alpha(\Delta_i)$. Next if $z_1 \in \Delta_i \setminus \mathbf{B}$ and $z_2 \in \mathbf{B} \setminus \Delta_i$, there is a point $w \in \mathbf{B} \cap \Delta_i$. By uniformity there are curves γ_1 and γ_2 joining z_1 and z_2 to w in Δ_i and $\mathbf{B}$, respectively, satisfying (3.5.2) and (3.5.3). By generalizing the argument in the proof of Theorem 6.2.4 it is easy to see that f lies in the right Lipschitz class. The last case when $z_1 \in \Delta_i \setminus \mathbf{B}$ and $z_2 \in \Delta_j \setminus \mathbf{B}$, $i \neq j$, follows similarly by joining z_1 and z_2 to $w_1 \in \Delta_i \cap \mathbf{B}$ and $w_2 \in \Delta_j \cap \mathbf{B}$, respectively, and then joining w_1 and w_2 by a third curve in $\mathbf{B}$.

We have, however, the following characterization of quasidisks in terms of the Hardy-Littlewood propety.

THEOREM 6.2.10 (Gehring-Martio [**64**]). *Suppose that D is a simply connected domain in $\mathbf{R}^2$ with $\infty \in \partial D$ and D^* a domain. Then D is a quasidisk if and only if D and D^* have the Hardy-Littlewood property.*

This result will follow immediately from Theorem 6.1.2 if we can establish that the Hardy-Littlewood property of some order implies property 1° in Definition 2.4.1. In the original paper [**64**] this was done via Lemma 9.3.1. We will indicate another route in the next section which sheds some light on the Hardy-Littlewood property and related notions.

6.3. Lip_α-extension domains

We return to Theorem 6.2.4. This result can be viewed as a consequence of two implications. First, if (6.2.5) holds in D, then by Theorem 6.2.3

$$|f(z_1) - f(z_2)| \leq \frac{c_0 m}{\alpha} |z_1 - z_2|^\alpha$$

in each open disk in D. Thus f belongs to the local Lipschitz class $\mathrm{loc\,Lip}_\alpha(D)$. Next, the fact that D is uniform implies that f is in $\mathrm{Lip}_\alpha(D)$ with an appropriate norm. In general a domain D with the property that there exists a constant $A = A(D, \alpha)$ such that

$$\|f\|_\alpha \leq A \, \|f\|_\alpha^{\mathrm{loc}}$$

for all real- or complex-valued functions f in the local Lipschitz class $\mathrm{loc\,Lip}_\alpha(D)$ is called a Lip_α*-extension domain.* This class of domains can be characterized by the following distance inequality proved in [**66**].

THEOREM 6.3.1 (Gehring-Martio [**66**]). *Suppose* $\alpha \in (0, 1]$. *A domain* D *in* $\mathbf{R}^2$ *is a* Lip_α*-extension domain if and only if there exists a constant* $M = M(D, \alpha)$ *such that for all* $z_1, z_2 \in D$

$$d_\alpha(z_1, z_2) = \inf_\gamma \int_\gamma \mathrm{dist}(z, \partial D)^{\alpha - 1} ds \leq M |z_1 - z_2|^\alpha \tag{6.3.2}$$

where γ *is a rectifiable curve joining* z_1 *and* z_2 *in* D.

Now let

$$\delta_\alpha(z_1, z_2) = \sup_f |f(z_1) - f(z_2)|,$$

where the supremum is taken over all analytic functions f on D satisfying (6.2.7). Then the metrics d_α introduced in (6.3.2) and δ_α are comparable. This is the content of the next important result due to Kaufman and Wu.

THEOREM 6.3.3 (Kaufman-Wu [**99**]). *If* D *is a simply connected domain in* $\mathbf{R}^2$, *then*

$$\delta_\alpha \leq d_\alpha \leq c_1 \, \delta_\alpha, \quad \alpha \in (0, 1], \tag{6.3.4}$$

where c_1 *is an absolute constant.*

It turns out that the simply connected Lip_α-extension domains for a fixed α are exactly the Hardy-Littlewood domains of order α.

THEOREM 6.3.5 (Astala-Hag-Hag-Lappalainen [**15**]). *A simply connected domain in* $\mathbf{R}^2$ *has the Hardy-Littlewood property of order* $\alpha \in (0, 1]$ *if and only if it is a* Lip_α*-extension domain.*

PROOF. Assume that D has the Hardy-Littlewood property of order α. Then

$$\delta_\alpha(z_1, z_2) \leq \frac{c(D)}{\alpha} |z_1 - z_2|^\alpha$$

and thus by (6.3.4)

$$d_\alpha(z_1, z_2) \leq \frac{c_1 c(D)}{\alpha} |z_1 - z_2|^\alpha,$$

which shows that condition (6.3.2) is satisfied and hence that D is a Lip_α-extension domain.

Next, assume that D is a Lip_α-extension domain, i.e. that (6.3.2) is satisfied. Consider f analytic in D satisfying (6.2.7). Then

$$|f(z_1) - f(z_2)| \le \delta_\alpha(z_1, z_2) \le d_\alpha(z_1, z_2) \le M|z_1 - z_2|^\alpha$$

and $\|f\|_\alpha \le M = M(D, \alpha)$. □

Now suppose that D is a Lip_α-extension domain for the class of analytic functions only. It follows from Theorem 6.2.3 that if an analytic function satisfies (6.2.2) in D, then it lies in $\mathrm{loc}\,\mathrm{Lip}_\alpha(D)$ and hence in $\mathrm{Lip}_\alpha(D)$. Thus D has the Hardy-Littlewood property of order α, and we obtain the following corollary to Theorem 6.3.5.

COROLLARY 6.3.6. *If a simply connected domain is a* Lip_α*-extension domain for the class of analytic functions, then it is a* Lip_α*-extension domain.*

Next we prove that the Lip_α-extension domains possess property 1° in Definition 2.4.1.

THEOREM 6.3.7 (Gehring-Martio [**66**]). *Suppose that a domain D is a* Lip_α*-extension domain. Then there is a constant $c \ge 1$ depending only on α and M in* (6.2.2) *such that for each $z_0 \in \mathbf{R}^2$ and each $r > 0$ points in $D \cap \mathbf{B}(z_0, r)$ can be joined in $D \cap \mathbf{B}(z_0, cr)$.*

PROOF. Choose $z_1, z_2 \in D \cap \mathbf{B}(z_0, r)$ where we may assume that $\partial D \cap \overline{\mathbf{B}}(z_0, r)$ contains some point w; otherwise $c = 1$ will work. By Theorem 6.3.1 there exists a curve γ joining z_1 and z_2 in D so that

$$\int_\gamma \mathrm{dist}(z, \partial D)^{\alpha-1} ds \le 2M|z_1 - z_2|^\alpha \le 2M(2r)^\alpha. \tag{6.3.8}$$

Suppose that γ is not contained in $D \cap \mathbf{B}(z_0, cr)$, where $c > 1$. Then

$$\begin{aligned}\int_\gamma \mathrm{dist}(z, \partial D)^{\alpha-1} ds &\ge \int_\gamma |z - w|^{\alpha-1} ds \ge \int_\gamma (|z - z_0| + r)^{\alpha-1} ds \\ &\ge 2\int_r^{cr} (t + r)^{\alpha-1} dt = \frac{2r^\alpha}{\alpha}\left((c+1)^\alpha - 2^\alpha\right).\end{aligned}$$

This, together with (6.3.8), yields a bound for c depending only on α and M. □

COROLLARY 6.3.9 (Lappalainen [**111**]). *If D is a* Lip_α*-extension domain and if $0 < \alpha \le \beta \le 1$, then D is also a* Lip_β*-extension domain.*

PROOF. Fix $z_1, z_2 \in D$ and choose $z_0 = (z_1 + z_2)/2$, $r = |z_1 - z_2|/2$. Then z_1 and z_2 can be joined by a curve γ in $\mathbf{B}(z_0, cr)$, $c = 2(M+1)^{1/\alpha}$, so that (6.3.8) is satisfied, where M is as in (6.3.8).

If $\overline{\mathbf{B}}(z_0, cr) \cap \partial D = \emptyset$, then

$$\int_{[z_1, z_2]} \mathrm{dist}(z, \partial D)^{\beta-1}\, ds \le r^{\beta-1} 2r = 2r^\beta \le 2|z_1 - z_2|^\beta.$$

If $\overline{\mathbf{B}}(z_0, cr) \cap \partial D \ne \emptyset$, then $\mathrm{dist}(z, \partial D) \le 2cr$ for every $z \in \gamma \subset \mathbf{B}(z_0, cr)$ and

$$2M(2r)^\alpha \ge \int_\gamma \mathrm{dist}(z, \partial D)^{\alpha-1}\, ds \ge (2cr)^{\alpha-\beta} \int_\gamma \mathrm{dist}(z, \partial D)^{\beta-1}\, ds,$$

i.e. $\displaystyle\int_\gamma \mathrm{dist}(z, \partial D)^{\beta-1}\, ds \le 2c^{\beta-\alpha} M|z_1 - z_2|^\beta.$ □

Finally, we have a characterization of quasidisks in terms of Lip_α-extension domains parallel to Theorem 6.2.10.

THEOREM 6.3.10 (Gehring-Martio [**66**]). *Suppose D is a simply connected domain in $\mathbf{R}^2$ with $\infty \in \partial D$ and D^* a domain. Then D is a quasidisk if and only if D is a Lip_α-extension domain and D^* is a Lip_β-extension domain with $\alpha, \beta \in (0,1]$.*

This result, as well as Theorem 6.2.10, follows from the fact that the domains D and D^* possess property 1° in Definition 2.4.1 according to Theorem 6.3.7 and Theorem 6.1.2 applies.

6.4. Harmonic doubling condition

In Sections 3.8 and 3.9 we saw how quasidisks can be described by conditions involving harmonic measures. Here we will give yet another characterization in terms of harmonic measure, but this time the condition has to be satisfied simultaneously in D and D^*.

DEFINITION 6.4.1. Let $D \subset \overline{\mathbf{R}}^2$ be a Jordan domain with bounded boundary ∂D. We say that D satisfies a *harmonic doubling condition*, or simply that D is *doubling*, if there exists a point $z_0 \in D$ and a constant $b \geq 1$ such that if γ_1 and γ_2 are adjacent arcs in ∂D with

$$\mathrm{diam}(\gamma_2) \leq \mathrm{diam}(\gamma_1),$$

then

$$\omega(z_0, \gamma_2; D) \leq b\,\omega(z_0, \gamma_1; D).$$

The inequality for the diameters may be replaced by an equality in the above definition.

The doubling condition is clearly satisfied in $\mathbf{B}$ and $\mathbf{B}^*$, but not in $\mathbf{H}$ since (3.7.2) shows that in this case boundary arcs of large Euclidean diameters need not carry much harmonic measure. Hence it is necessary to restrict our attention to domains whose boundaries lie entirely in $\mathbf{R}^2$.

REMARK 6.4.2. A bounded Jordan domain D is a disk if and only if D and D^* both satisfy a harmonic doubling condition with constant $b = 1$.

PROOF. To establish sufficiency, suppose that D and D^* satisfy a harmonic doubling condition for $b = 1$ and the points $z_0 \in D, z_0^* \in D^*$, and let γ_1 and γ_2 be consecutive boundary arcs with $\mathrm{diam}(\gamma_1) = \mathrm{diam}(\gamma_2)$. Then $\omega(z_0, \gamma_1; D) = \omega(z_0, \gamma_2; D)$ and $\omega(z_0^*, \gamma_1; D^*) = \omega(z_0^*, \gamma_2; D^*)$. It is not difficult to see that

$$\omega(z_0, \gamma_1; D) = \omega(z_0, \gamma_2; D) \quad \text{implies} \quad \mathrm{diam}(\gamma_1) = \mathrm{diam}(\gamma_2), \tag{6.4.3}$$

and the same conclusion holds in D^* as well.

Let $\phi\colon \mathbf{B} \to D$ and $\psi\colon \mathbf{B}^* \to D^*$ be conformal maps with $\phi(0) = z_0$ and $\psi(\infty) = z_0^*$. Let $I_1, I_2 \subset \mathbf{S}^1 = \partial\mathbf{B}$ be consecutive arcs with $\mathrm{length}(I_1) = \mathrm{length}(I_2)$ or, equivalently, $\omega(0, I_1; \mathbf{B}) = \omega(0, I_2; \mathbf{B})$. Then by conformal invariance $\omega(z_0, \phi(I_1); D) = \omega(z_0, \phi(I_2); D)$, and (6.4.3) implies that $\mathrm{diam}(\phi(I_1)) = \mathrm{diam}(\phi(I_2))$. The doubling condition yields

$$\omega(z_0^*, \phi(I_1); D^*) = \omega(z_0^*, \phi(I_2); D^*),$$

so that by conformal invariance once more we obtain

$$\omega(\infty, f(I_1); \mathbf{B}) = \omega(\infty, f(I_2); \mathbf{B}),$$

where $f = \psi^{-1} \circ \phi \colon \mathbf{S}^1 \to \mathbf{S}^1$ denotes the sewing homeomorphism associated with D. We have proved that

$$\text{(6.4.4)} \qquad \text{length}(I_1) = \text{length}(I_2) \quad \text{implies} \quad \text{length}(f(I_1)) = \text{length}(f(I_2)).$$

By performing two rotations, if necessary, we may assume that $f(1) = 1$. Then by continuity (6.4.4) implies that $f(-1) = -1$. Continuing as in the proof of Theorem 3.8.1, we see that f fixes every point $e^{i\theta}$ on $\mathbf{S}^1$ with $\theta = m\pi/2^n$, $n = 1, 2, \ldots$ and $m = 1, \ldots, 2 \cdot 2^n$. By continuity f is the identity mapping on $\mathbf{S}^1$. Thus D must be a disk as in the proof of Theorem 3.8.1. □

A harmonic doubling condition in a Jordan domain is not enough to guarantee that domain to be a quasidisk. In fact the doubling Jordan domains are exactly the John disks.

THEOREM 6.4.5 (Kim [**101**], Kim and Langmeyer [**102**]). *A bounded Jordan domain D in $\mathbf{R}^2$ is doubling if and only if it is a John disk.*

The next result, proved by Jerison and Kenig [**89**], characterizes quasidisks in terms of the doubling condition for harmonic measure. We will follow the proof given in Broch-Hag-Junge [**29**]. See also Garnett-Marshall [**45**].

THEOREM 6.4.6 (Jerison and Kenig [**89**]). *A bounded Jordan domain D in $\mathbf{R}^2$ is a quasidisk if and only if D and D^* both are doubling.*

PROOF. For sufficiency, suppose that D and D^* satisfy the harmonic doubling condition with respect to $z_0 \in D$ and $z_0^* \in D^*$, with the same constant b. We will show that then D is a quasisymmetric domain and hence a quasidisk by Theorem 3.8.8. To this end let γ_1 and γ_2 be adjacent arcs of ∂D with

$$\text{(6.4.7)} \qquad \omega(z_0, \gamma_1; D) = \omega(z_0, \gamma_2; D).$$

Without loss of generality we may assume that

$$\text{(6.4.8)} \qquad \text{diam}(\gamma_1) \le \text{diam}(\gamma_2).$$

In the following we will use the notation $\omega(\gamma) = \omega(z_0, \gamma; D)$ and $\omega^* = \omega(z_0^*, \gamma; D^*)$.

Denote by z_1 and z_2 the endpoints of γ_1 and by z_2 and z_3 the endpoints of γ_2. For $z, w \in \gamma_1 \cup \gamma_2$ we let $\gamma(z, w)$ denote the arc of ∂D from z to w contained in $\gamma_1 \cup \gamma_2$ and order points along $\gamma_1 \cup \gamma_2$ in the direction from z_1 to z_3.

We choose $w_0, w_1, \ldots, w_n$ on γ_2 as follows. Let $w_0 = z_2$. Suppose that points $w_0, w_1, \ldots, w_k$ have been chosen along $\gamma_2 = \gamma(z_2, z_3)$ so that for all $0 \le i \le k-1$, w_{i+1} is the first point with

$$\text{diam}(\gamma(w_k, w_{k+1})) = \text{diam}(\gamma(z_1, w_k)).$$

If $\text{diam}(\gamma(w_k, z_3)) < \text{diam}(\gamma(w_k, z_1))$, then we stop and set $n = k$. Otherwise, continue. This process must stop with $n \le b$, for by using the doubling condition in D we have that

$$\omega(\gamma_1) \le \omega(\gamma(z_1, w_k)) \le b\,\omega(\gamma(w_k, w_{k+1})).$$

This implies that

$$n\,\omega(\gamma_1) \le b \sum_{k=0}^{n-1} \omega(\gamma(w_k, w_{k+1})) \le b\,\omega(\gamma_2),$$

and by (6.4.7) we must conclude that $n \le b$.

Next we will use the doubling condition in D^*. For this we choose $n \leq b$ as above so that $\operatorname{diam}(\gamma(w_n, z_3)) < \operatorname{diam}(\gamma(z_1, w_n))$ and $\omega(\gamma(w_n, z_3)) \leq b\,\omega(\gamma(z_1, w_n))$. Hence

$$\begin{aligned}\omega(\gamma(z_2, z_3)) &= \omega(\gamma(z_2, w_n)) + \omega(\gamma(w_n, z_3)) \\ &\leq (1+b)\,\omega(\gamma(z_1, w_n)).\end{aligned} \tag{6.4.9}$$

Moreover, for $k = 0, 1, \ldots, n-1$ we have that

$$\operatorname{diam}(\gamma(w_k, w_{k+1})) = \operatorname{diam}(\gamma(z_1, w_k))$$

and

$$\omega^*(\gamma(w_k, w_{k+1})) \leq b\,\omega^*(\gamma(z_1, w_k)).$$

From this we see that

$$\begin{aligned}\omega^*(\gamma(z_1, w_{k+1})) &= \omega^*(\gamma(z_1, w_k)) + \omega^*(\gamma(w_k, w_{k+1})) \\ &\leq (1+b)\,\omega^*(\gamma(z_1, w_k)),\end{aligned}$$

for $k = 0, 1, \ldots, n-1$. Combining (6.4.9) with these n inequalities we get

$$\begin{aligned}\omega^*(\gamma(z_2, z_3)) &\leq (b+1)^{n+1}\,\omega^*(\gamma(z_1, z_2)) \\ &\leq (b+1)^{b+1}\,\omega^*(\gamma(z_1, z_2)).\end{aligned}$$

All in all we have that

$$\frac{1}{b}\,\omega^*(\gamma_1) \leq \omega^*(\gamma_2) \leq (b+1)^{b+1}\omega^*(\gamma_1)$$

and D is quasisymmetric with $c = (b+1)^{b+1}$.

Conversely, assume that D is a bounded quasidisk and let $\phi\colon D \to \mathbf{B}$ be a conformal mapping. Then ϕ has an extension to a quasiconformal self-mapping of $\overline{\mathbf{R}}^2$ which we also denote by ϕ. See Corollary 2.1.5. By composing ϕ with a Möbius transformation, if necessary, we may assume that $\phi(\infty) = \infty$. Let γ_1 and γ_2 be adjacent arcs in ∂D with $\operatorname{diam}(\gamma_1) = \operatorname{diam}(\gamma_2)$ and let the pairs z_1, z_2 and z_2, z_3 be their endpoints, respectively. By Ahlfors' three-point condition (Theorem 2.2.5) we have that

$$|z_3 - z_2| \leq \operatorname{diam}(\gamma_2) = \operatorname{diam}(\gamma_1) \leq a|z_2 - z_1|,$$

where $a = a(D)$. By the distortion result in Corollary 1.3.7 we obtain

$$|\phi(z_3) - \phi(z_2)| \leq c|\phi(z_2) - \phi(z_1)|,$$

$c = c(a)$. In the following we assume without loss of generality that the arc $\phi(\gamma_2)$ is the larger arc. If this arc subtends an angle of less than π at the origin, then

$$\omega(0, \phi(\gamma_2); \mathbf{B}) \leq c\frac{\pi}{2}\,\omega(0, \phi(\gamma_1); \mathbf{B}).$$

In the case when $\phi(\gamma_2)$ subtends an angle larger than π we look at the subarc of $\phi(\gamma_2)$ from $\phi(z_2)$ to $\phi(z_3')$ of length π and apply Ahlfors' three-point condition to $\gamma_2' = \gamma(z_2, z_3')$ and $\gamma_1' = \gamma(z_1', z_2)$ of the same diameter and proceed as before. In this case we have that

$$\omega(0, \phi(\gamma_2); \mathbf{B}) \leq \pi c\,\omega(0, \phi(\gamma_1); \mathbf{B}),$$

and we conclude that D is doubling. In the same way we prove that D^* is doubling. □

CHAPTER 7

Miscellaneous properties

We conclude our account of ways to describe a quasidisk D with some seemingly unrelated characterizations. These include the continuity of Bloch functions, homogeneity properties of D and ∂D, Dirichlet integrals of functions harmonic in D and D^*, a description of the family of all quasicircles, and finally a relation between quasiconformal mappings in $\overline{\mathbf{R}}^2$ and in $\overline{\mathbf{R}}^3$.

7.1. Bloch functions

A function f analytic in $D \subset \mathbf{R}^2$ is said be a *Bloch* function, or is in $B(D)$, if

$$\|f\|_{B(D)} = \sup_{z \in D} |f'(z)| \operatorname{dist}(z, \partial D) < \infty. \tag{7.1.1}$$

Bloch functions play an important role in complex analysis. For example, if

$$f(z) = \log g'(z) \tag{7.1.2}$$

where g is conformal in a simply connected domain $D \subset \mathbf{R}^2$, then

$$f'(z) = T_g(z)$$

where T_g is the pre-Schwarzian derivative of g. Hence f is in $B(D)$ with

$$\|f\|_{B(D)} = \sup_{z \in D} |T_g(z)| \operatorname{dist}(z, \partial D) \leq 2 \sup_{z \in D} |T_g(z)|\, \rho_D(z)^{-1} \leq 8$$

by inequality (3.2.1) and Theorem 4.1.11. For other examples see Bonk [**26**] and Liu-Minda [**119**].

The bound for $f'(z)$ in (7.1.1) implies that functions f in $B(D)$ have the following modulus of continuity when D is a disk or half-plane.

EXAMPLE 7.1.3. If f is in $B(D)$ where D is a disk or half-plane, then

$$|f(z_1) - f(z_2)| \leq \|f\|_{B(D)}\, j_D(z_1, z_2)$$

for $z_1, z_2 \in D$, where j_D is the distance-ratio metric defined in (3.3.2).

Note that if D is a disk or half-plane, then

$$|f'(z)| \leq \|f\|_{B(D)} \frac{1}{\operatorname{dist}(z, \partial D)} \leq \|f\|_{B(D)}\, \rho_D(z)$$

for $z \in D$ and

$$|f(z_1) - f(z_2)| \leq \|f\|_{B(D)}\, h_D(z_1, z_2) \leq \|f\|_{B(D)}\, j_D(z_1, z_2)$$

by Example 3.3.7.

This result does not hold for all domains D. For example if

$$D = \{z = x + i\, y : 0 < x < \infty, |y| < 1\}$$

and $f(z) = z$, then $g(z) = e^z$ is conformal in D, $f(z) = \log g'(z)$, and

$$\|f\|_{B(D)} \leq 8$$

by what was proved above. However when $x > 1$,

$$\frac{|f(x) - f(1)|}{j_D(x,1)} = \frac{|x-1|}{2\,\log|x|} \to \infty$$

as $x \to \infty$.

The simply connected domains D where Bloch functions f have a multiple of the metric j_D as a modulus of continuity are quasidisks.

THEOREM 7.1.4 (Langmeyer [**110**]). *A simply connected domain $D \subset \mathbf{R}^2$ is a quasidisk if and only if there exists a constant $a > 0$ such that*

$$|f(z_1) - f(z_2)| \leq a\, \|f\|_{B(D)}\, j_D(z_1, z_2)$$

for each f in $B(D)$ and $z_1, z_2 \in D$.

The above result suggests the following question. For what kinds of conformal mappings $g(z)$ are the functions in (7.1.2) Bloch functions? The answer yields another characterization for quasidisks.

THEOREM 7.1.5 (Broch [**28**]). *A simply connected domain $D \subset \mathbf{R}^2$ is a quasidisk if and only if there exists a constant $b > 0$ such that each f in $B(D)$ with $\|f\|_{B(D)} \leq b$ is of the form*

$$f(z) = \log g'(z)$$

where g is conformal in D.

PROOF. Suppose that f is analytic in D and let

$$g(z) = \int_{z_0}^{z} e^{f(\zeta)} d\zeta$$

where $z_0 \in D$. Then g is analytic with

$$f(z) = \log g'(z), \qquad f'(z) = \frac{g''(z)}{g'(z)} = T_g(z)$$

for $z \in D$ and

$$\frac{1}{2} \sup_D |T_g(z)| \rho_D(z)^{-1} \leq \|f\|_{B(D)} \leq 2 \sup_D |T_g(z)| \rho_D(z)^{-1}$$

by (3.2.1) and (7.1.1).

If D is a quasidisk, then by Theorem 4.1.14 there exists a constant $c > 0$ such that g is conformal in D whenever

$$\sup_D |T_g(z)| \rho_D(z)^{-1} \leq 2c$$

and hence whenever $\|f\|_{B(D)} \leq c$.

Conversely if there exists a constant $c > 0$ such that $f = \log g'$ is in $B(D)$ with g conformal in D whenever $\|f\|_{B(D)} \leq c$ and hence whenever

$$\sup_D |T_g(z)| \rho_D(z)^{-1} \leq \frac{c}{2},$$

then D is a quasidisk by Theorem 4.1.14. □

7.2. Comparable Dirichlet integrals

We observed in Theorem 5.2.3 that if $D \subset \mathbf{R}^2$ is a simply connected domain, then each function u in $L^2_1(D)$ can be extended to a function v in $L^2_1(\mathbf{R}^2)$ if and only if D is a quasidisk. We consider here a similar problem, namely to determine the Jordan domains $D \subset \mathbf{R}^2$ for which functions harmonic in D and D^* with equal boundary values on ∂D have comparable Dirichlet integrals.

If D is a disk or half-plane and if u and v are harmonic in D and D^*, respectively, with continuous and equal boundary values, then

$$\int_D |\text{grad } u|^2 dm = \int_{D^*} |\text{grad } v|^2 dm. \tag{7.2.1}$$

The converse is also true. If D is a Jordan domain and if (7.2.1) holds for each pair of functions u and v harmonic in D and D^* with continuous and equal boundary values, then D is a disk or half-plane (Hag [**77**]).

The following example shows what we should expect if D is a quasidisk.

EXAMPLE 7.2.2. If D is a sector of angle α, then there exist functions u and v harmonic in D and D^*, respectively, with continuous and equal boundary values such that

$$\int_{D^*} |\text{grad } v(z)|^2\, dm = \frac{2\pi - \alpha + \sin\alpha}{\alpha - \sin\alpha} \int_D |\text{grad } u(z)|^2\, dm. \tag{7.2.3}$$

For a proof of this suppose that $D = \mathbf{S}(\alpha)$ and let

$$f(z) = \frac{1-z}{1+z}, \qquad u(z) = \text{Re}(f(z))$$

for $z \in \mathbf{R}^2 \setminus \{-1\}$ and let

$$g(z) = \frac{1+z}{1-z}, \qquad v(z) = \text{Re}(g(\bar{z}))$$

for $z \in \mathbf{R}^2 \setminus \{1\}$. Then u is harmonic in D, v is harmonic in D^*, and $u(z) = v(z)$ for $z \in \partial D \setminus \{\infty\}$.

Next $f(D) = f(\mathbf{S}(\alpha))$ is a domain that contains the interval $I = (-1,1)$ and is bounded by two circular arcs which meet I at angles $\pm\alpha/2$ at its endpoints. An elementary calculation shows that

$$m(f(D)) = \frac{\alpha - \sin\alpha}{\sin^2(\alpha/2)}. \tag{7.2.4}$$

Similarly $g(D^*) = f(\mathbf{S}(2\pi - \alpha))$ is bounded by circular arcs which meet I at angles $\pm(\pi - \alpha/2)$ at its endpoints and

$$m(g(D^*)) = \frac{2\pi - \alpha + \sin\alpha}{\sin^2(\alpha/2)} \tag{7.2.5}$$

as above. Then

$$\int_D |\text{grad } u(z)|^2\, dm = \int_D |f'(z)|^2\, dm = m(f(D)),$$

$$\int_{D^*} |\text{grad } v(z)|^2\, dm = \int_{D^*} |g'(z)|^2\, dm = m(g(D^*))$$

and we obtain (7.2.3) from (7.2.4) and (7.2.5).

DEFINITION 7.2.6. A Jordan domain D has the *comparable Dirichlet integral property* if there exists a constant $c \geq 1$ such that

$$\frac{1}{c}\int_D |\text{grad }u|^2 dm \leq \int_{D^*} |\text{grad }v|^2 dm \leq c\int_D |\text{grad }u|^2 dm$$

for each pair of functions u and v which are harmonic in D and D^*, respectively, with continuous and equal boundary values.

THEOREM 7.2.7 (Ahlfors [**4**], Springer [**154**]). *A Jordan domain D is a K-quasidisk if and only if D has the comparable Dirichlet integral property with constant c, where K and c depend only on each other.*

7.3. Quasiconformal groups

Suppose that G is a group of self-homeomorphisms g of $\overline{\mathbf{R}}^2$, i.e., a family which is closed under composition and taking inverses. We say that w_0 is in the *limit set* $L(G)$ of G if there exist distinct $g_j \in G$ and a point $z_0 \in \overline{\mathbf{R}}^2$ such that

$$w_0 = \lim_{j\to\infty} g_j(z_0). \tag{7.3.1}$$

Suppose next that D is a K-quasidisk. Then there exists a K-quasiconformal self-mapping f of $\mathbf{R}^2$ that maps the upper half-plane $\mathbf{H}$ onto D. Let g_1 and g_2 be the Möbius transformations

$$g_1(z) = -1/z, \qquad g_2(z) = z + 1,$$

and let $G = \langle g_1, g_2\rangle$, the group generated by g_1 and g_2. Then G has $\partial\mathbf{H}$ as its limit set (Beardon [**18**]). Next let

$$F = \{f \circ g \circ f^{-1} : g \in G\}.$$

Then F is a group of K^2-quasiconformal self-mappings of $\overline{\mathbf{R}}^2$.

If $w_1 \in \partial D$, then $w_0 = f^{-1}(w_1) \in \partial\mathbf{H}$ and there exists a sequence of distinct $g_j \in G$ and a point $z_0 \in \overline{\mathbf{R}}^2$ for which (7.3.1) holds. Thus

$$w_1 = \lim_{j\to\infty} f \circ g_j \circ f^{-1}(z_1) \in L(F)$$

where $z_1 = f(z_0)$. Reversing the above argument shows that each w_1 in $L(F)$ is a point of ∂D.

We conclude that the boundary of each quasidisk is the limit set of a finitely generated group of self-mappings of $\overline{\mathbf{R}}^2$ that are K-quasiconformal for some fixed K. The following result shows that this property actually characterizes the family of quasidisks.

THEOREM 7.3.2 (Maskit [**125**], Sullivan [**156**], Tukia [**161**]). *A Jordan domain D is a quasidisk if and only if ∂D is the limit set of a finitely generated group of K-quasiconformal self-mappings of $\overline{\mathbf{R}}^2$ for some fixed K.*

7.4. Homogeneity

We recall the standing assumption that D is a subdomain of $\overline{\mathbf{R}}^2$ of hyperbolic type.

A set $E \subset \overline{\mathbf{R}}^2$ is *homogeneous* with respect to a family $\mathcal{F}$ of mappings if for each $z_1, z_2 \in E$ there exists a mapping $f \in \mathcal{F}$ such that

$$f(E) = E, \qquad f(z_1) = z_2.$$

Every simply connected domain $D \subset \overline{\mathbf{R}}^2$ is homogeneous with respect to conformal, and hence quasiconformal, mappings of D onto itself. The situation changes if we consider self-mappings of $\overline{\mathbf{R}}^2$. For example, disks and half-planes are the only homogeneous Jordan domains with respect to conformal mappings of $\overline{\mathbf{R}}^2$. This result has a counterpart for quasidisks and quasiconformal mappings.

We let $QC(K)$ denote the family of all K-quasiconformal self-mappings of $\overline{\mathbf{R}}^2$. Then $QC(1)$ is simply the family of mappings generated by the family $\mathcal{M}$ and the reflection $r(z) = \overline{z}$.

If D is a K-quasidisk, then $D = f(\mathbf{H})$ and $\partial D = f(\partial \mathbf{H})$ where f is a K-quasiconformal self-mapping of $\overline{\mathbf{R}}^2$. Hence D and ∂D are *both* homogeneous with respect to the family of mappings

$$\mathcal{F} = f \circ \mathcal{M} \circ f^{-1} \subset QC(K^2).$$

Moreover a Jordan domain D is a quasidisk if *either* D or its boundary ∂D is homogeneous with respect to the family $QC(K)$ for some K. More precisely we have the following results.

THEOREM 7.4.1 (Brechner, Erkama [**27**], [**37**]). *A simply connected domain D is a quasidisk if and only if ∂D is homogeneous with respect to the family $QC(K)$ for some fixed K.*

THEOREM 7.4.2 (Sarvas [**153**]). *A Jordan domain D is a quasidisk if and only if it is homogeneous with respect to the family $QC(K)$ for some fixed K.*

The following example shows that the hypothesis that D be a Jordan domain is necessary in Theorem 7.4.2.

EXAMPLE 7.4.3 (Hjelle [**86**]). There exists a simply connected domain D which is not a quasidisk but which is homogeneous with respect to the family $QC(K)$ for a fixed $K > 1$.

We will sketch a proof of this. Let G denote the group of 1-quasiconformal mappings generated by the reflections g_1, g_2 in the circles $|z+2| = 1$, $|z-2| = 1$ and the translation

$$h(z) = z + i$$

and let G_0 denote the family consisting of the identity and the sixteen transformations

$$g_j, \qquad h, \qquad h^{-1}, \qquad g_j \circ h, \qquad g_j \circ h^{-1}, \qquad h \circ g_j, \qquad h^{-1} \circ g_j,$$
$$h^{-1} \circ g_j \circ h, \qquad h \circ g_j \circ h^{-1}$$

where $j = 1, 2$. Next let

$$E = \{z = x + i\,y : |x| \le 2 - \sqrt{1 - y^2}, \ \ |y| \le 1/2\},$$

and let

$$D = \bigcup_{g \in G} g(E), \qquad D_0 = \operatorname{int}\Big(\bigcup_{g \in G_0} g(E)\Big).$$

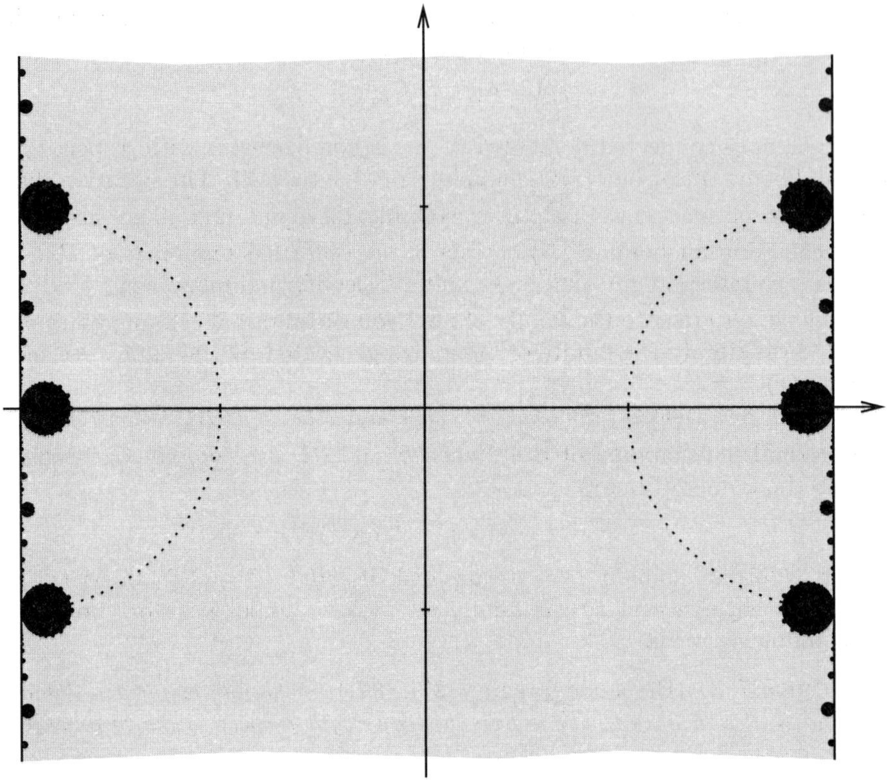

FIGURE 7.1

Hjelle shows that D is the union of an increasing sequence of simply connected domains which lie in $\{z = x + i\,y : |x| < 2\}$. Hence D is itself a simply connected domain. See, for example, page 81 in Newman [**140**].

Next D_0 is a Jordan domain which contains E. A theorem due to Teichmüller implies that for each pair of points $z_1, z_2 \in E$ there exists a K-quasiconformal mapping $f : \mathbf{R}^2 \to \mathbf{R}^2$ such that

1° $f(z_1) = z_2$,
2° $f(z) = z$ for $z \in \mathbf{R}^2 \setminus D_0$,

where K depends only on $h_{D_0}(z_1, z_2)$. See Teichmüller [**158**]. Thus D is homogeneous with respect to the family

$$\mathcal{F} = G \circ \{f\} \circ G \subset QC(K).$$

However D is not locally connected at $\infty \in \partial D$. Hence D is not a Jordan domain and, in particular, not a quasidisk.

On the other hand, the hypothesis that D be a Jordan domain in Theorem 7.4.2 is not needed when $K = 1$.

THEOREM 7.4.4 (Erkama [**38**], Kimel′fel′d [**103**]). *A simply connected domain D is a disk or half-plane if and only if it is homogeneous with respect to the family $QC(1)$.*

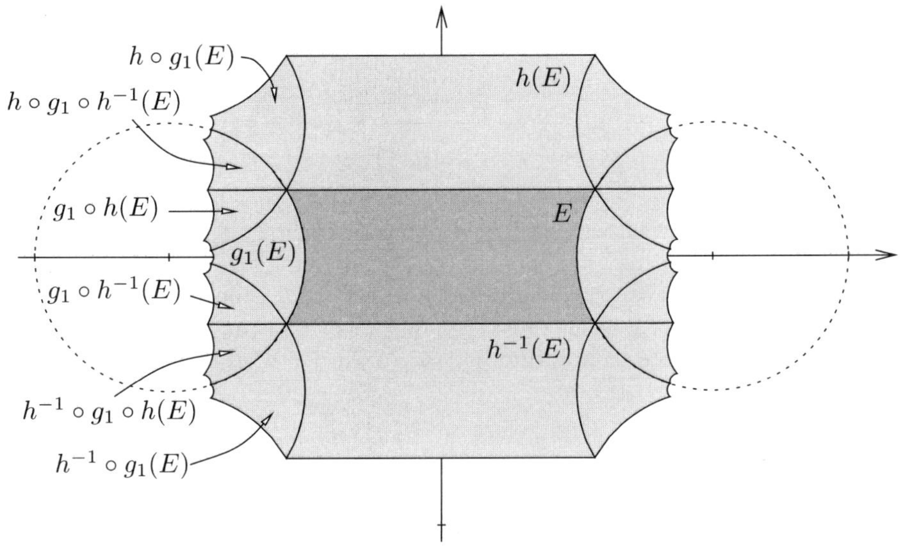

FIGURE 7.2

7.5. Family of all quasicircles

We construct next a simple family Σ of Jordan curves $S \subset \mathbf{R}^2$ such that the boundary ∂D of each bounded quasidisk D is bilipschitz equivalent to a curve S in Σ. The construction is a generalization of the construction of von Koch's well-known snowflake curve.

Given r with $1/4 \le r < 1/2$, we set

$$s = \sqrt{r - 1/4}$$

and assign to each segment $\gamma = [a, b] \subset \mathbf{R}^2$ a polygonal curve $\gamma(r)$ consisting of the union of four segments

$$\begin{aligned} \gamma_1 &= [a,\ a + r(b-a)], \\ \gamma_2 &= [a + r(b-a),\ \tfrac{1}{2}(a+b) - i\,s(b-a)], \\ \gamma_3 &= [\tfrac{1}{2}(a+b) - i\,s(b-a),\ b - r(b-a)], \\ \gamma_4 &= [b - r(b-a),\ b]. \end{aligned} \tag{7.5.1}$$

Hence $\operatorname{length}(\gamma_j) = r \operatorname{length}(\gamma)$ for $j = 1, 2, 3, 4$ and

1° $\gamma(r)$ consists of four equal segments in γ when $r = 1/4$,
2° $\gamma(r)$ consists of two segments in γ plus a "triangle" with base in γ on the right side of γ when $1/4 < r < 1/2$.

Now for each parameter $1/4 < p < 1/2$ we define a family $\Sigma(p)$ of polygons S_n as follows. First let S_1 denote the positively oriented unit square with sides

$$\gamma_1 = [0, 1], \qquad \gamma_2 = [1, 1+i], \qquad \gamma_3 = [1+i, i], \qquad \gamma_4 = [i, 0].$$

Then let S_2 be any of the 2^4 polygons obtained by replacing each of the four sides γ_j of S_1 by the polygonal curve

$$\gamma_j(r) = \bigcup_{k=1}^{4} \gamma_{j,k} \quad \text{for } r = 1/4 \text{ or } r = p,$$

where each of $\gamma_{j,1}, \gamma_{j,2}, \gamma_{j,3}$, and $\gamma_{j,4}$ is as described in (7.5.1).

Continuing in this way, we pass from any one of the $2^{4^{n-1}}$ polygons S_n to one of the corresponding 2^{4^n} new polygons S_{n+1} by replacing each of the 4^n sides γ of S_n by the polygonal curve $\gamma(r)$ where $r = 1/4$ or $r = p$. Since each point of S_{n+1} lies within distance p^n of a point of S_n, every sequence of polygons S_n obtained in this manner will converge to a Jordan curve S. We let $\Sigma(p)$ denote the collection of all such limit curves S and set

$$\Sigma = \bigcup_{1/4<p<1/2} \Sigma(p).$$

We then have the following description of the bounded quasidisks in $\mathbf{R}^2$.

THEOREM 7.5.2 (Rohde [**151**]). *A bounded domain $D \subset \mathbf{R}^2$ is a quasidisk if and only if there exist a curve $S \in \Sigma$ and a bilipschitz mapping $f : \mathbf{R}^2 \to \mathbf{R}^2$ such that*

$$\partial D = f(S).$$

7.6. Quasiconformal equivalence of $\overline{\mathbf{R}}^3 \setminus \overline{D}$ and $\mathbf{B}^3$

We conclude our list of descriptions by showing how one can characterize two-dimensional quasidisks in terms of their properties as subsets of three-dimensional space. We begin observing that Definition 1.1.3 can be generalized ad verbatim to n-space.

Suppose that D and D' are domains in $\overline{\mathbf{R}}^n$ and that $f : D \to D'$ is a homeomorphism. For $x \in D \setminus \{\infty, f^{-1}(\infty)\}$ and $0 < r < \text{dist}(x, \partial D)$ we let

$$l_f(x,r) = \min_{|x-y|=r} |f(x) - f(y)|, \qquad L_f(x,r) = \max_{|x-y|=r} |f(x) - f(y)|$$

and call

$$H_f(x) = \limsup_{r \to 0} \frac{L_f(x,r)}{l_f(x,r)}$$

the *linear dilatation* of f at x.

A homeomorphism $f : D \to D'$ is *K-quasiconformal* where $1 \le K < \infty$ if $H_f(x) < \infty$ for every $x \in D \setminus \{\infty, f^{-1}(\infty)\}$ and

$$H_f(x) \le K$$

almost everywhere in D.

EXAMPLE 7.6.1. If D is a sector of angle α in $\mathbf{R}^2$, then $\overline{\mathbf{R}}^3 \setminus \overline{D}$ can be mapped K-quasiconformally onto the unit ball $\mathbf{B}^3$ in $\mathbf{R}^3$ where

$$K = 2 \max\left(\frac{\pi}{\alpha}, \frac{\pi}{2\pi - \alpha}\right).$$

To see this, we may assume without loss of generality that $D = \mathbf{S}(\alpha)$ where $0 < \alpha \leq \pi$. Let $h : \overline{\mathbf{R}}^3 \to \overline{\mathbf{R}}^3$ be given in cylindrical coordinates $(r, \theta, x_3) \in \mathbf{R}^3$ by

$$h(r, \theta, x_3) = (r, \phi(\theta), x_3)$$

and $h(\infty) = \infty$, where

$$\phi(\theta) = \begin{cases} \dfrac{\pi\,\theta}{\alpha} & \text{if } 0 \leq \theta \leq \dfrac{\alpha}{2}, \\ \pi - \dfrac{\pi\,(\pi - \theta)}{2\,\pi - \alpha} & \text{if } \dfrac{\alpha}{2} \leq \theta \leq \pi \end{cases}$$

and $\phi(-\theta) = -\phi(\theta)$. Then h is $K/2$-quasiconformal, where K is as above, and h maps D onto the real half-plane

$$\mathbf{H}^2 = \{(x_1, x_2, x_3) : 0 < x_1 < \infty,\ |x_2| < \infty,\ x_3 = 0\}$$

and hence $\overline{\mathbf{R}}^3 \setminus \overline{D}$ onto $\overline{\mathbf{R}}^3 \setminus \overline{\mathbf{H}}^2$.

Next there exists a 2-quasiconformal mapping g which unfolds $\overline{\mathbf{R}}^3 \setminus \overline{\mathbf{H}}^2$ around the x_2-axis onto the upper half-space

$$\mathbf{H}^3 = \{(x_1, x_2, x_3) \in \mathbf{R}^3 : |x_1| < \infty,\ |x_2| < \infty,\ 0 < x_3 < \infty\}.$$

Finally, let f be a Möbius transformation which carries $\mathbf{H}^3$ onto $\mathbf{B}^3$. Then $f \circ g \circ h$ is K-quasiconformal and maps $\overline{\mathbf{R}}^3 \setminus \overline{D}$ onto $\mathbf{B}^3$.

The problem of determining whether or not a domain $D \subset \mathbf{R}^n$ can be mapped onto the unit ball $\mathbf{B}^n$ is quite difficult when $n \geq 3$. However the property of linear local connectivity yields the following necessary condition.

EXAMPLE 7.6.2 (Gehring-Väisälä [**70**]). If $D \subset \mathbf{R}^n$ can be mapped K-quasiconformally onto $\mathbf{B}^n$, then there exists a constant $c = c(K)$ such that for each $x_0 \in \mathbf{R}^n$ and each $r > 0$

$$E \cap \mathbf{B}^n(x_0, r) \ \text{ lies in a component of } \ E \cap \mathbf{B}^n(x_0, cr),$$
$$E \setminus \overline{\mathbf{B}}^n(x_0, r) \ \text{ lies in a component of } \ E \setminus \overline{\mathbf{B}}^n(x_0, r/c),$$

where $E = \overline{\mathbf{R}}^n \setminus D$.

These examples illustrate the following interesting relation between quasiconformal mappings in $\mathbf{R}^2$ and $\mathbf{R}^3$.

THEOREM 7.6.3 (Gehring [**48**]). *A domain $D \subset \mathbf{R}^2$ is a quasidisk if and only if $\overline{\mathbf{R}}^3 \setminus \overline{D}$ can be mapped quasiconformally onto $\mathbf{B}^3$.*

SKETCH OF PROOF. If D is a quasidisk, a construction similar to that in Example 7.6.1 plus an important lifting theorem for quasiconformal mappings (Ahlfors [**6**]) shows that $\overline{\mathbf{R}}^3 \setminus \overline{D}$ can be mapped quasiconformally onto $\mathbf{B}^3$.

If $\overline{\mathbf{R}}^3 \setminus D$ can be mapped quasiconformally onto $\mathbf{B}^3$, then Example 7.6.2 with $n = 3$ implies that D is linearly locally connected and hence a quasidisk by Theorem 2.4.4. □

Part 2

Some proofs of these properties

CHAPTER 8

First series of implications

In the preceding chapters we have presented many different ways to view a quasidisk D in $\mathbf{R}^2$. We turn now to the proofs of some of these characterizations. Our first goal is to prove the following statements for simply connected domains D with $\operatorname{card}(\overline{\mathbf{R}}^2 \setminus D) \geq 2$.

1° A quasidisk D has the hyperbolic segment property.
2° The hyperbolic segment property implies D is uniform.
3° A uniform domain is linearly locally connected.
4° Linear local connectivity implies the three-point condition.
5° The three-point condition implies the quadrilateral inequality.
6° The quadrilateral inequality implies D is a quasidisk.
7° D is a quasidisk if and only if it admits reflections.
8° D is a quasidisk if and only if it is quasiconformally decomposable.

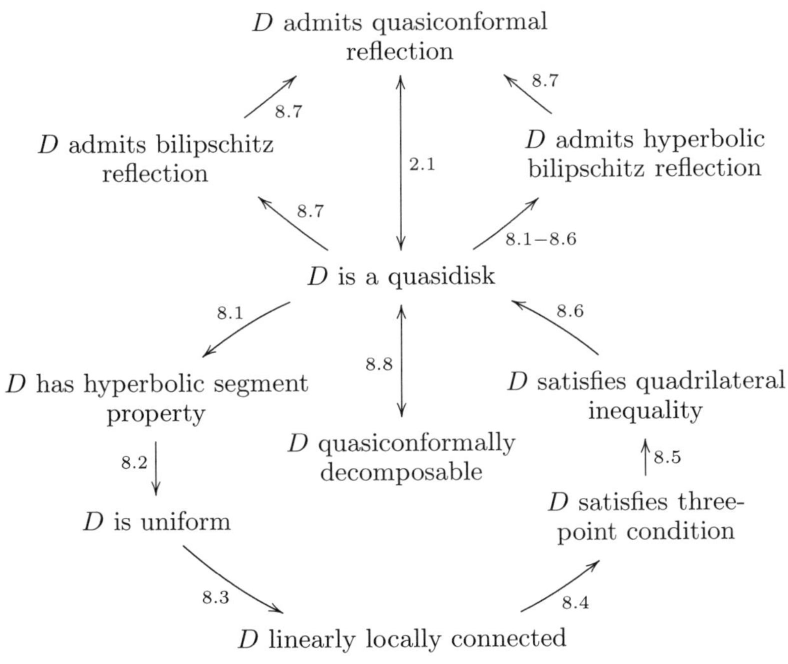

FIGURE 8.1

8.1. Quasidisks and hyperbolic segments

We begin with the following elementary observation.

LEMMA 8.1.1. *Suppose that D is a disk or half-plane, that D' is a K-quasidisk, and that $f : D \to D'$ is conformal. Then f has an extension which is K^2-quasiconformal in $\overline{\mathbf{R}}^2$.*

PROOF. By hypothesis there exists a K-quasiconformal mapping $g : \overline{\mathbf{R}}^2 \to \overline{\mathbf{R}}^2$ with $g(D') = \mathbf{B}$. Then $h = g \circ f$ is a K-quasiconformal mapping of D onto $\mathbf{B}$ which extends as a homeomorphism of $\overline{D}$ onto $\overline{\mathbf{B}}$ by Theorem 1.3.11. Let ϕ and ψ denote the reflections in ∂D and $\partial \mathbf{B}$, respectively, and set

$$h(z) = \psi \circ g \circ f \circ \phi^{-1}(z)$$

for $z \in D^*$. Then h is K-quasiconformal in D, in D^*, and hence in $\overline{\mathbf{R}}^2$ by Theorem 1.3.12. Set

$$f(z) = g^{-1} \circ h(z)$$

for $z \in D^*$. Then f is K^2-quasiconformal in D^*, in D, and hence in $\overline{\mathbf{R}}^2$. □

The next result will allow us to estimate the length of a hyperbolic segment. The proof is based on an argument due to Jerison and Kenig [**88**].

LEMMA 8.1.2. *Suppose that D is a K-quasidisk with $\infty \in \partial D$ and that $f : \overline{\mathbf{H}} \to \overline{D}$ is a homeomorphism which is conformal in $\mathbf{H}$ with $f(\infty) = \infty$. Then*

$$\int_0^y |f'(i\,t)|\,dt \le c_1 \operatorname{dist}(f(i\,y), \partial D) \tag{8.1.3}$$

for $0 < y < \infty$ where $c_1 = c_1(K)$.

PROOF. By Lemma 8.1.1, f has an extension which is K^2-quasiconformal in $\overline{\mathbf{R}}^2$. By a change of variables we may assume that $f(0) = 0$. Fix $0 < y_0 < \infty$ and choose a sequence $\{y_j\}$ so that

$$0 < y_{j+1} < y_j \le y_0$$

and

$$|f(i\,y_j)| = c_2^{-j}\,|f(i\,y_0)|$$

for $j = 1, 2, \ldots$, where $c_2 = e^{8K^2}$. Then for $y_{j+1} \le y \le y_j$,

$$\operatorname{dist}(f(i\,y), \partial D) \le |f(i\,y) - f(0)| \le c_2\,|f(i\,y_j) - f(0)| = c_2^{-j+1}\,|f(i\,y_0)|$$

by Theorem 1.3.4 since $f(0) = 0$, while

$$|f'(i\,y)| \le 2\,\frac{\operatorname{dist}(f(i\,y), \partial D)}{\operatorname{dist}(i\,y, \partial \mathbf{H})} \le 2\,c_2^{-j+1}\frac{|f(i\,y_0)|}{y}$$

by the Koebe distortion theorem. Thus

$$\int_{y_{j+1}}^{y_j} |f'(i\,y)|\,dy \le 2\,c_2^{-j+1}|f(i\,y_0)|\,\log\frac{y_j}{y_{j+1}}\,. \tag{8.1.4}$$

Next let k denote the smallest positive integer for which $c_2 \le 2^k$. Then

$$|f(i\,y_j) - f(0)| \le 2^k\,|f(i\,y_{j+1}) - f(0)|,$$

and hence

$$y_j = |iy_j - 0| \le (2c_2)^{k+1}\,|iy_{j+1} - 0| = (2c_2)^{k+1}\,y_{j+1}$$

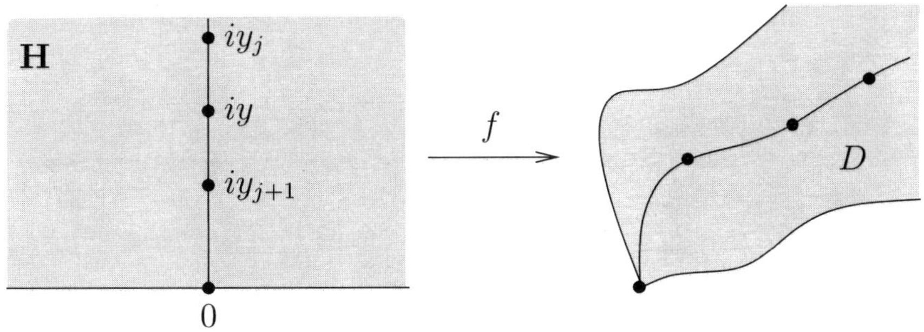

FIGURE 8.2

by Corollary 1.3.7. This yields

$$\log \frac{y_j}{y_{j+1}} \leq (k+1) \log 2c_2 = c_3\,,$$

and with (8.1.4) we obtain

$$\begin{aligned}\int_0^{y_0} |f'(i\,y)|\,dy &= \sum_{j=0}^{\infty} \int_{y_{j+1}}^{y_j} |f'(i\,y)|\,dy \\ &\leq 2c_3 |f(i\,y_0)| \sum_{j=0}^{\infty} c_2^{-j+1} = c_4 |f(i\,y_0)|.\end{aligned}$$

Finally, if $x \in \partial\mathbf{H}$, then

$$|f(i\,y_0)| \leq c_2\,|f(i\,y_0) - f(x)|$$

by Theorem 1.3.4 again and thus

$$|f(i\,y_0)| \leq c_2 \operatorname{dist}(f(i\,y_0), \partial\mathbf{H}).$$

This completes the proof of (8.1.3) with $c_1 = c_2\,c_4$, a constant which depends only on K. □

THEOREM 8.1.5 (Gehring-Osgood [**67**]). *Suppose that γ is a hyperbolic segment joining z_1 and z_2 in a K-quasidisk $D \subset \mathbf{R}^2$. Then there exists a constant $c = c(K)$ such that for each $z \in \gamma$,*

(8.1.6) $$\operatorname{length}(\gamma) \leq c\,|z_1 - z_2|,$$

(8.1.7) $$\min_{j=1,2} \operatorname{length}(\gamma_j) \leq c \operatorname{dist}(z, \partial D),$$

where γ_1, γ_2 are the components of $\gamma \setminus \{z\}$.

PROOF. By Lemma 8.1.1 there exists a K^2-quasiconformal self-mapping f of $\overline{\mathbf{R}}^2$ which maps D conformally onto $\mathbf{B}$. By employing an auxiliary Möbius transformation of the disk we may assume that $f(z_1)$ and $f(z_2)$ are real. Next let B' be the open disk in $\mathbf{B}$ with $f(z_1)$ and $f(z_2)$ as diametral points. Then $D' = f^{-1}(B')$ is a bounded K^2-quasidisk and γ is a hyperbolic line in D'. Since

$$\operatorname{dist}(z, \partial D') \leq \operatorname{dist}(z, \partial D)$$

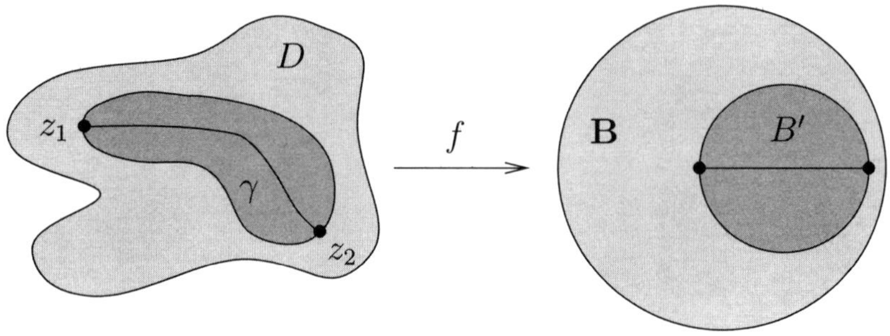

FIGURE 8.3

for $z \in \gamma$, we see that it is sufficient to establish (8.1.6) and (8.1.7) for the case where γ is a hyperbolic line in D and D is bounded.

Assume now that this is the case and let D' and γ' be the images of D and γ under

$$w = g(z) = \frac{z - z_1}{z - z_2} .$$

Let f map $\mathbf{H}$ conformally onto D'. Again f extends to a homeomorphism of $\overline{\mathbf{H}}$ onto $\overline{D'}$ and we may assume that $f(0) = 0$ and $f(\infty) = \infty$.

Now D' is a K-quasidisk and γ' is the image of the positive imaginary axis under f. Hence if $w \in \gamma'$, then $w = f(i\,y)$ for some $0 < y < \infty$ and

$$s = \int_0^y |f'(it)|\, dt \leq c_1 \operatorname{dist}(f(i\,y), \partial D') = c_1 \operatorname{dist}(w, \partial D') \tag{8.1.8}$$

by Lemma 8.1.2 where s denotes the arclength of γ' between 0 and w. Let $h = g^{-1}$. Then

$$\begin{aligned} \operatorname{length}(\gamma) &= \int_{\gamma'} |h'(w)|\, |dw| = |z_1 - z_2| \int_{\gamma'} \frac{|dw|}{|w-1|^2} \\ &= |z_1 - z_2| \int_0^\infty \frac{ds}{|w(s) - 1|^2} , \end{aligned}$$

where $w = w(s)$ is the arclength representation for γ'. If we let

$$s_0 = \frac{c_1}{c_1 + 1} ,$$

then for $0 < s \leq s_0$,

$$|w(s) - 1| \geq 1 - |w(s)| \geq 1 - s \geq \frac{1}{c_1 + 1} ,$$

while for $s_0 \leq s < \infty$,

$$|w(s) - 1| \geq \operatorname{dist}(w(s), \partial D') \geq \frac{s}{c_1}$$

by (8.1.8) since $D \subset \mathbf{R}^2$ implies that

$$1 = g(\infty) \notin D'.$$

Thus

$$\int_0^\infty \frac{ds}{|w(s) - 1|^2} \leq \int_0^{s_0} (c_1 + 1)^2\, ds + \int_{s_0}^\infty \frac{c_1^2}{s^2}\, ds = 2c_1(c_1 + 1) = c_2$$

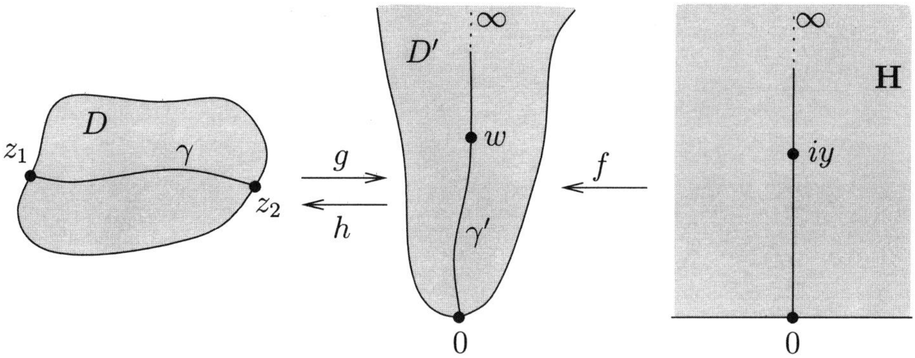

FIGURE 8.4

and we obtain (8.1.6) with $c = c_2$.

Finally, since D is bounded, we can find a K^2-quasiconformal self-mapping f of $\overline{\mathbf{R}}^2$ which maps D conformally onto $\mathbf{B}$ with $f(\infty) = \infty$. Fix $z \in \gamma$, choose $z_0 \in \partial D$ so that

$$|z - z_0| = \operatorname{dist}(z, \partial D),$$

and let w_1, w_2, w, and w_0 be the images of z_1, z_2, z, and z_0 under f.

Since $f(\gamma)$ is a hyperbolic line in $\mathbf{B}$, it is easy to check that

$$\min_{j=1,2} |w - w_j| \leq 2 \operatorname{dist}(w, \partial \mathbf{B}) \leq 2\,|w - w_0|$$

and hence

$$\min_{j=1,2} |z_j - z| \leq 2c_3^2\,|z - z_0| = 2c_3^2 \operatorname{dist}(z, \partial D)$$

by Corollary 1.3.7, where $c_3 = e^{8K^2}$. If γ_j is the component of $\gamma \setminus \{z\}$ which has z_j as an endpoint, then

$$\min_{j=1,2} \operatorname{length}(\gamma_j) \leq c_2\,|z_j - z|$$

and we obtain (8.1.7) with $c = 2\,c_2\,c_3^2$. □

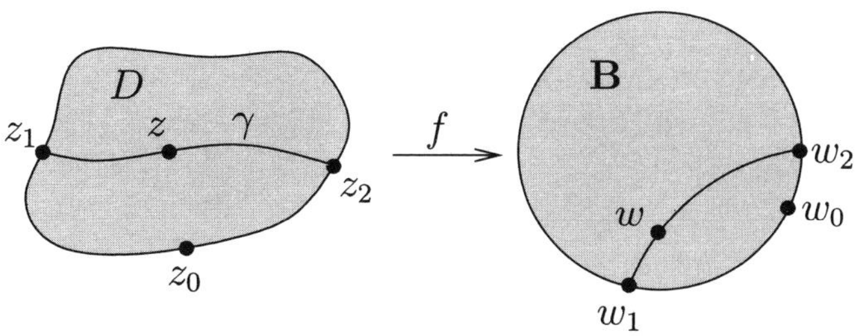

FIGURE 8.5

8.2. Hyperbolic segments and uniform domains

The following is an immediate consequence of the earlier Definition 3.5.1.

REMARK 8.2.1. Suppose that D is a simply connected domain in $\mathbf{R}^2$ and that there is a constant $c \geq 1$ such that for each hyperbolic segment γ joining z_1 and z_2 and each $z \in \gamma$,

$$\text{length}(\gamma) \leq c\,|z_1 - z_2|, \tag{8.2.2}$$

$$\min_{j=1,2} \text{length}(\gamma_j) \leq c\,\text{dist}(z, \partial D), \tag{8.2.3}$$

where γ_1, γ_2 are the components of $\gamma \setminus \{z\}$. Then D is a uniform domain with constant c.

Thus a simply connected domain with the hyperbolic segment property is uniform.

8.3. Uniform domains and linear local connectivity

We establish next a rather general result to show that uniform domains are linearly locally connected.

THEOREM 8.3.1. *Suppose that D is a domain in $\mathbf{R}^2$ and that for some constants $a, b \geq 1$ each pair of points z_1, z_2 in D can be joined by arcs α and β in D such that*

$$\text{diam}(\alpha) \leq a\,|z_1 - z_2|, \tag{8.3.2}$$

$$\min_{j=1,2} |z - z_j| \leq b\,\text{dist}(z, \partial D) \tag{8.3.3}$$

for $z \in \beta$. Then for each $z_0 \in \mathbf{R}^2$ and each $r > 0$

$$D \cap \mathbf{B}(z_0, r) \text{ lies in a component of } D \cap \mathbf{B}(z_0, c\,r), \tag{8.3.4}$$

$$D \setminus \overline{\mathbf{B}}(z_0, r) \text{ lies in a component of } D \setminus \overline{\mathbf{B}}(z_0, r/c), \tag{8.3.5}$$

where $c = 2\max(a, b) + 1$.

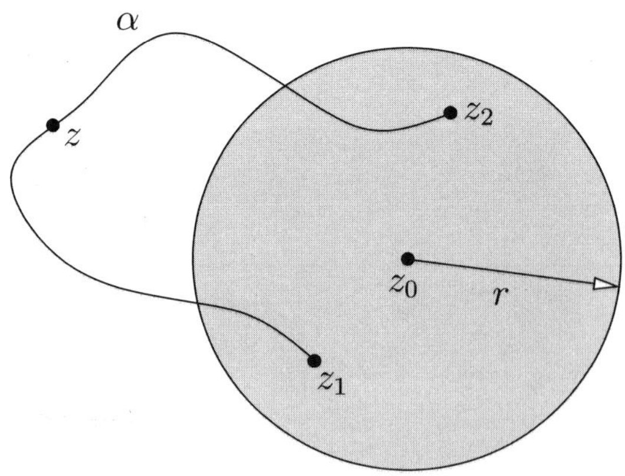

FIGURE 8.6

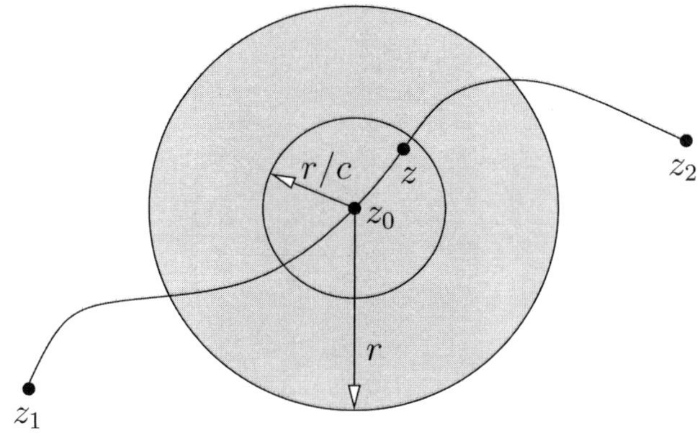

FIGURE 8.7

PROOF. Fix $z_0 \in \mathbf{R}^2$, $0 < r < \infty$, and suppose that $z_1, z_2 \in D \cap \mathbf{B}(z_0, r)$. We must show that z_1 and z_2 can be joined in $D \cap \mathbf{B}(z_0, c\,r)$. By hypothesis there exists an arc α joining z_1 and z_2 in D with

$$\operatorname{diam}(\alpha) \le a\,|z_1 - z_2| \le 2a\,r.$$

If $z \in \alpha$, then

$$|z - z_0| \le |z - z_1| + |z_1 - z_0| < \operatorname{diam}(\alpha) + r \le (2\,a + 1)r \le c\,r,$$

and hence α joins z_1 and z_2 in $D \cap \mathbf{B}(z_0, c\,r)$ as required.

Suppose next that $z_1, z_2 \in D \setminus \overline{\mathbf{B}}(z_0, r)$. Again by hypothesis there exists an arc β joining z_1 and z_2 in D with

$$\min_{j=1,2} |z - z_j| \le b\,\operatorname{dist}(z, \partial D)$$

for each $z \in \beta$. Suppose

$$\beta \not\subset D \setminus \overline{\mathbf{B}}(z_0, r/c).$$

Then there exists a point $z \in \beta$ with

$$|z - z_0| < \frac{r}{c} \le \frac{r}{2b + 1}\,,$$

and for $j = 1, 2$ we have

$$|z_j - z| \ge |z_j - z_0| - |z - z_0| > \frac{2b}{2b + 1}\,r.$$

Thus

$$\operatorname{dist}(z, \partial D) \ge \frac{1}{b} \min_{j=1,2} |z - z_j| > \frac{2}{2b + 1}\,r \ge |z - z_0| + \frac{r}{c}$$

and hence $\overline{\mathbf{B}}(z_0, r/c) \subset D$. But this implies that $D \setminus \overline{\mathbf{B}}(z_0, r/c)$ is connected and hence that z_1, z_2 can be joined by an arc in this set. Thus z_1, z_2 can always be joined in $D \setminus \overline{\mathbf{B}}(z_0, r/c)$. □

If D is a uniform domain in $\mathbf{R}^2$, then D satisfies the hypotheses in Theorem 8.3.1 with $\alpha = \beta$ and $a = b$. Hence uniform domains are linearly locally connected.

We shall see in the following three sections that a simply connected domain which is linearly locally connected is a quasidisk and hence, by what has gone

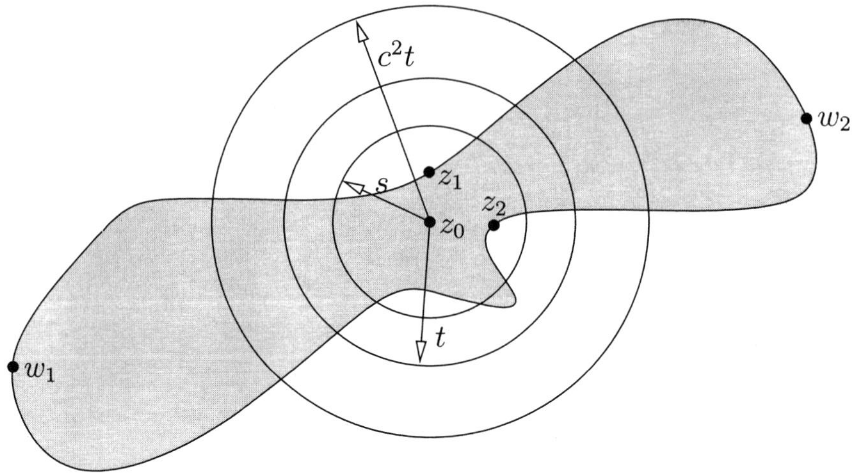

FIGURE 8.8

before, uniform. Thus conditions (3.5.2) and (3.5.3) in Definition 3.5.1 for a uniform domain can be replaced by the substantially weaker hypotheses (8.3.2) and (8.3.3) when D is simply connected.

8.4. Linear local connectivity and the three-point condition

We show next that a simply connected domain which is linearly locally connected satisfies the three-point condition.

THEOREM 8.4.1 (Gehring [**49**]). *Suppose that D is a simply connected domain in $\mathbf{R}^2$ and that there exists a constant $c \geq 1$ such that for each $z_0 \in \mathbf{R}^2$ and $r > 0$*

(8.4.2) $$D \cap \mathbf{B}(z_0, r) \text{ lies in a component of } D \cap \mathbf{B}(z_0, cr),$$

(8.4.3) $$D \setminus \overline{\mathbf{B}}(z_0, r) \text{ lies in a component of } D \setminus \overline{\mathbf{B}}(z_0, r/c).$$

Then D is a Jordan domain and for each pair of points $z_1, z_2 \in \partial D \setminus \{\infty\}$

(8.4.4) $$\min_{j=1,2} \operatorname{diam}(\gamma_j) \leq c^2 \, |z_1 - z_2|$$

where γ_1, γ_2 are the components of $\partial D \setminus \{z_1, z_2\}$.

PROOF. The above hypotheses imply that D is locally connected at each point of its boundary and hence a Jordan domain by the converse of the Jordan curve theorem (Newman [**140**]).

Next suppose that (8.4.4) does not hold for two points $z_1, z_2 \in \partial D \setminus \{\infty\}$ and set

$$z_0 = \frac{1}{2}\,(z_1 + z_2), \qquad r = \frac{1}{2}\,|z_1 - z_2|.$$

Then there exist t with $r < t < \infty$ and finite points w_1, w_2 such that

$$w_j \in \gamma_j \setminus \mathbf{B}(z_0, c^2\, t)$$

for $j = 1, 2$.

Choose s with $r < s < t$. Since

$$z_1, z_2 \in \partial D \cap \mathbf{B}(z_0, s),$$

we can find for $j = 1, 2$ an endcut α_j of D which joins z_j to a point $z_j' \in D$ and which lies in $D \cap \mathbf{B}(z_0, s)$. Next by (8.4.2) we can find an arc α_3 joining z_1' and z_2' in $D \cap \mathbf{B}(z_0, cs)$. Then $\alpha_1 \cup \alpha_2 \cup \alpha_3$ contains a crosscut α of D which joins z_1, z_2 in $\mathbf{B}(z_0, cs)$.

The same argument applied to w_1, w_2 yields a crosscut β of D which joins w_1, w_2 in $\overline{D} \setminus \overline{\mathbf{B}}(z_0, ct)$. Then

$$\alpha \cap \beta \subset \mathbf{B}(z_0, cs) \setminus \overline{\mathbf{B}}(z_0, ct) = \emptyset$$

since $s < t$, while the fact that z_1, z_2 separate w_1, w_2 in ∂D implies that $\alpha \cap \beta \neq \emptyset$ and we have a contradiction. □

8.5. The three-point condition and quadrilaterals

We show next that the three-point condition for a Jordan domain D implies an important relation between conjugate quadrilaterals in D and its exterior D^*.

LEMMA 8.5.1. *Suppose that $D \subset \mathbf{R}^2$ is a Jordan domain and that b is a constant such that for each $z_1, z_2 \in \partial D \setminus \{\infty\}$*

$$\min_{j=1,2} \operatorname{diam}(\gamma_j) \leq b\,|z_1 - z_2| \tag{8.5.2}$$

where γ_1, γ_2 are the components of $\partial D \setminus \{z_1, z_2\}$. If Q and Q^* are conjugate quadrilaterals in D and D^* and if $\operatorname{mod}(Q) = 1$, then

$$\operatorname{mod}(Q^*) \leq c, \tag{8.5.3}$$

where $c = c(b)$.

PROOF. Let α_1 and α_2 denote two opposite sides of Q and Q^* and let Γ and Γ^* denote the family of arcs which join α_1 and α_2 in Q and Q^*, respectively. Then

$$\operatorname{mod}(Q^*) = \operatorname{mod}(\Gamma^*), \qquad \operatorname{mod}(\Gamma) = \operatorname{mod}(Q) = 1$$

and hence it suffices to shows that

$$\operatorname{mod}(\Gamma^*) \leq \pi(2\,b^2\,e^{2\pi} + 1)^2. \tag{8.5.4}$$

Choose $z_1 \in \alpha_1$ and $z_2 \in \alpha_2$ so that

$$|z_1 - z_2| = \operatorname{dist}(\alpha_1, \alpha_2) = r$$

and let

$$s = \min_{j=1,2} \operatorname{diam}(\alpha_j) < \infty.$$

We show first that

$$\frac{s}{r} \leq 2\,b^2 e^{2\pi}. \tag{8.5.5}$$

We may assume that $2b^2\,r < s$ since otherwise (8.5.5) follows trivially.

Let γ_1, γ_2 denote the components of $\partial D \setminus \{z_1, z_2\}$. Then by (8.5.2) and relabeling, if necessary, we may assume that

$$\operatorname{diam}(\gamma_1) = \min_{j=1,2} \operatorname{diam}(\gamma_j) \leq b|z_1 - z_2| = b\,r.$$

Next let β_1, β_2 denote the components of $\partial D \setminus (\alpha_1 \cup \alpha_2)$, labeled so that $\beta_j \subset \gamma_j$ and choose $w_1 \in \beta_1$ and $w_2 \in \beta_2$. Then

$$|w_1 - z_1| \le \operatorname{diam}(\gamma_1) \le b|z_1 - z_2| = b\,r.$$

If δ_1, δ_2 denote the components of $\partial D \setminus \{w_1, w_2\}$, then

$$s = \min_{j=1,2} \operatorname{diam}(\alpha_j) \le \min_{j=1,2} \operatorname{diam}(\delta_j) \le b|w_1 - w_2|$$

by (8.5.2) and hence

$$|w_2 - z_1| \ge |w_2 - w_1| - |w_1 - z_1| \ge \frac{s}{b} - b\,r \ge \frac{s}{2\,b}.$$

Thus for $br < t < s/2b$ each circle $\{z : |z - z_1| = t\}$ intersects α_1 and α_2 and separates β_1 from β_2. Hence each such circle contains an arc γ which joins α_1 and α_2 in D, i.e., an arc $\gamma \in \Gamma$. Lemma 1.3.2 implies that

$$1 = \operatorname{mod}(\Gamma) \ge \frac{1}{2\pi} \log \frac{s}{2\,b^2\,r}$$

from which (8.5.5) follows.

Finally, by (8.5.5) and Lemma 1.3.3,

$$\operatorname{mod}(\Gamma^*) \le \pi \left(\frac{s}{r} + 1\right)^2 \le \pi \left(2\,b^2\,e^{2\pi} + 1\right)^2$$

as desired. □

8.6. Quadrilateral inequality and quasidisks

Suppose that D_1 and D_2 are Jordan domains in $\overline{\mathbf{R}}^2$ and that $f : D_1 \to D_2$ is a K-quasiconformal mapping. Then by Theorem 1.3.11, f has a homeomorphic extension which maps $\overline{D}_1$ onto $\overline{D}_2$. We begin here by studying the boundary correspondence $\phi : \partial D_1 \to \partial D_2$ induced by f.

Suppose $g_j : D_j \to \mathbf{H}$ is conformal for $j = 1, 2$. Then g_j has a homeomorphic extension to $\overline{D}_j$,

$$h = g_2 \circ f \circ g_1^{-1}$$

is a self-homeomorphism of $\overline{\mathbf{H}}$ which is K-quasiconformal in $\mathbf{H}$, and

$$\phi = g_2^{-1} \circ \psi \circ g_1$$

where ψ is the boundary correspondence induced by h. Hence in order to study the mapping ϕ modulo conformal mappings, it is sufficient to consider the case where $D_1 = D_2 = \mathbf{H}$ and where $\phi(\infty) = \infty$.

We have next an important characterization due to Beurling and Ahlfors [**23**] for the boundary correspondences $\phi : \partial\mathbf{H} \to \partial\mathbf{H}$ induced by quasiconformal self-mappings f of $\mathbf{H}$. We then indicate why each such mapping ϕ is the boundary correspondence for a quasiconformal mapping f which is, in addition, bilipschitz with respect to the hyperbolic metric (Ahlfors [**7**]).

LEMMA 8.6.1. *Suppose that $\phi : \mathbf{R}^1 \to \mathbf{R}^1$ is the boundary correspondence induced by a K-quasiconformal mapping $f : \mathbf{H} \to \mathbf{H}$. Then ϕ is a homeomorphism and*

$$\frac{1}{k} \le \frac{\phi(x+t) - \phi(x)}{\phi(x) - \phi(x-t)} \le k \tag{8.6.2}$$

for all real x and $t > 0$, where k is a constant which depends only on K.

PROOF. Set $f(z) = \overline{f(\bar{z})}$ for $z \in \mathbf{H}^*$. Then f is a self-homeomorphism of $\mathbf{R}^2$ which is K-quasiconformal in $\mathbf{H}$, in $\mathbf{H}^*$, and hence in $\mathbf{R}^2$ by Theorem 1.3.12. Next $|(x+t)-x| = |x-(x-t)|$ and

$$|f(x+t)-f(x)| \le k\,|f(x)-f(x-t)|,$$
$$|f(x)-f(x-t)| \le k\,|f(x+t)-f(x)|$$

by Theorem 1.3.4 from which we obtain (8.6.2) with $k = e^{8K}$. □

The sharp estimate for the constant k in (8.6.2) is $k = \lambda(K)$ where

$$\lambda(K) = \left(\frac{1}{4}\,e^{\pi K/2} - e^{-\pi K/2}\right)^2 + \delta(K), \qquad 0 < \delta(K) < e^{-\pi K},$$

as in (1.3.6).

Functions $\phi : I \subset \mathbf{R}^1 \to \mathbf{R}^1$ which satisfy (8.6.2) for relevant x and t are said to be *k-quasisymmetric* (Kelingos [**100**]). They are, in a sense, one-dimensional quasiconformal mappings. On the real line this notion of quasisymmetry is equivalent to a condition which in general is strictly stronger and which can be used to define quasisymmetric mappings between metric spaces. See Heinonen [**80**] and also Astala-Iwaniec-Martin [**16**].

Lemma 8.6.1 implies that the boundary mapping induced by a quasiconformal self-mapping of $\mathbf{H}$ is quasisymmetric. The following important result due to Beurling and Ahlfors [**23**] shows that the converse is true. See also Ahlfors [**6**], Lehto [**116**], and Lehto-Virtanen [**117**].

THEOREM 8.6.3. *Suppose that $\phi : \mathbf{R}^1 \to \mathbf{R}^1$ is a homeomorphism and that*

$$\frac{1}{k} \le \frac{\phi(x+t)-\phi(x)}{\phi(x)-\phi(x-t)} \le k \tag{8.6.4}$$

for all real x and $t > 0$. Then there exists a K-quasiconformal mapping $f : \overline{\mathbf{R}}^2 \to \overline{\mathbf{R}}^2$ with

$$f(x) = \phi(x) \tag{8.6.5}$$

for $x \in \mathbf{R}^1$ where $K = K(k)$. Moreover f is L-bilipschitz with respect to the hyperbolic metric in $\mathbf{H}$, i.e.,

$$\frac{1}{L}\,h_{\mathbf{H}}(z_1,z_2) \le h_{\mathbf{H}}(f(z_1),f(z_2)) \le L\,h_{\mathbf{H}}(z_1,z_2) \tag{8.6.6}$$

for $z_1, z_2 \in \mathbf{H}$ where $L = L(k)$.

SKETCH OF PROOF. We may assume without loss of generality that $\phi(t)$ is increasing in t. For $z = x + iy \in \mathbf{R}^2$ set

$$f(z) = u(z) + i\,v(z) = \frac{1}{2}(\alpha(z)+\beta(z)) + \frac{i}{2}(\alpha(z)-\beta(z))$$

where

$$\alpha(z) = \int_0^1 \phi(x+ty)\,dt \qquad \text{and} \qquad \beta(z) = \int_0^1 \phi(x-ty)\,dt.$$

Then a technical argument based on inequality (8.6.4) shows that f is a homeomorphism and that

$$H_f(z) = \frac{1+|\mu_f(z)|}{1-|\mu_f(z)|} \le K$$

for $z \in \mathbf{H} \cup \mathbf{H}^*$ where

$$K = 2k(k+1).$$

See, for example, Beurling-Ahlfors [**23**], Lehto-Virtanen [**117**]. Hence f is K-quasiconformal by Theorem 1.3.12. A further calculation then yields (8.6.6) for $z_1, z_2 \in \mathbf{H}$ where

$$L = 4k^2(k+1).$$

See Ahlfors [**6**] for the details. □

We now use Theorem 8.6.3 to show that a Jordan domain D is a quasidisk if for all pairs of conjugate quadrilaterals Q and Q^* with boundary in ∂D, $\mathrm{mod}(Q^*)$ is bounded whenever $\mathrm{mod}(Q) = 1$.

THEOREM 8.6.7. *Suppose D is a Jordan domain in $\overline{\mathbf{R}}^2$ and suppose there exists a constant $c \geq 1$ such that*

$$\mathrm{mod}(Q^*) \leq c \tag{8.6.8}$$

whenever Q and Q^* are conjugate quadrilaterals with boundary in ∂D and

$$\mathrm{mod}(Q) = 1.$$

Then there exists a K-quasiconformal mapping $h : \overline{\mathbf{R}}^2 \to \overline{\mathbf{R}}^2$ such that $D = h(\mathbf{H})$ and such that

$$f^*(z) = h \circ r \circ h^{-1}(z) \qquad \text{where} \qquad r(z) = \bar{z} \tag{8.6.9}$$

defines a hyperbolic L-bilipschitz reflection in ∂D where $K = K(c)$ and $L = L(c)$.

PROOF. Choose conformal mappings $g : \mathbf{H} \to D$ and $g^* : \mathbf{H}^* \to D^*$. Then g and g^* have homeomorphic extensions to $\overline{\mathbf{H}}$ and $\overline{\mathbf{H}}^*$ and, by means of an auxiliary Möbius transformation, we may arrange that $g(\infty) = g^*(\infty)$. Hence

$$\phi(z) = g^{*-1} \circ g(z)$$

is a self-homeomorphism of $\mathbf{R}^1$ with $\phi(\infty) = \infty$.

Fix $x \in \mathbf{R}$ and $t > 0$ and let

$$\begin{aligned} z_1 &= x - t, & z_2 &= x, & z_3 &= x + t, & z_4 &= \infty, \\ w_1 &= \phi(x-t), & w_2 &= \phi(x), & w_3 &= \phi(x+t), & w_4 &= \infty. \end{aligned}$$

Then $Q_1 = \mathbf{H}(z_1, z_2, z_3, z_4)$ is a quadrilateral in $\mathbf{H}$,

$$[z_1, z_2, z_3, z_4] = \frac{z_3 - z_1}{z_3 - z_2} = 2,$$

and hence

$$\mathrm{mod}(Q_1) = \frac{2}{\pi}\mu\left(\frac{1}{\sqrt{2}}\right) = 1 \tag{8.6.10}$$

by (3.10.5) and (3.10.6). Similarly $Q_1^* = \mathbf{H}^*(w_4, w_3, w_2, w_1)$ is a quadrilateral in $\mathbf{H}^*$,

$$[w_4, w_3, w_2, w_1] = \frac{w_3 - w_1}{w_3 - w_2} = \frac{\phi(x+t) - \phi(x-t)}{\phi(x+t) - \phi(x)} = 1 + \frac{1}{u},$$

where

$$u = \frac{\phi(x+t) - \phi(x)}{\phi(x) - \phi(x-t)} \in (0, \infty),$$

and thus

$$\operatorname{mod}(Q_1^*) = \frac{2}{\pi}\mu\left(\frac{1}{\sqrt{1+1/u}}\right). \tag{8.6.11}$$

Let $Q = g(Q_1)$ and $Q^* = g^*(Q_1^*)$. Then Q and Q^* are conjugate quadrilaterals in D and D^* and

$$\operatorname{mod}(Q) = \operatorname{mod}(Q_1) = 1$$

by (8.6.10). Hence

$$\operatorname{mod}(Q^*) = \operatorname{mod}(Q_1^*) \le c$$

by (8.6.8) and

$$\frac{2}{\pi}\mu\left(\frac{1}{\sqrt{1+1/u}}\right) \le c$$

by (8.6.11).

The above argument, with Q_1 and Q_1^* replaced by $Q_2 = \mathbf{H}(z_2, z_3, z_4, z_1)$ and $Q_2^* = \mathbf{H}^*(w_1, w_4, w_3, w_2)$, implies that

$$\frac{2}{\pi}\mu\left(\frac{1}{\sqrt{1+u}}\right) \le c.$$

Then since

$$\lim_{r\to 0}\mu(r) = \infty,$$

there exists a constant $k = k(c) \ge 1$ such that

$$\frac{1}{k} \le \frac{\phi(x+t) - \phi(x)}{\phi(x) - \phi(x-t)} \le k. \tag{8.6.12}$$

Now (8.6.12) holds for all $x \in \mathbf{R}$ and $t > 0$. Hence by Theorem 8.6.3 there exists a K-quasiconformal mapping $f : \overline{\mathbf{R}}^2 \to \overline{\mathbf{R}}^2$ where $K = K(k)$ such that $f(\mathbf{H}) = \mathbf{H}$ and $f(x) = \phi(x)$ for $x \in \partial\mathbf{H}$. In addition,

$$\frac{1}{L}\,h_{\mathbf{H}}(w_1, w_2) \le h_{\mathbf{H}}(f(w_1), f(w_2)) \le L\,h_{\mathbf{H}}(w_1, w_2) \tag{8.6.13}$$

for $w_1, w_2 \in \mathbf{H}$ where $L = L(k)$. Then

$$h(z) = \begin{cases} g \circ f^{-1}(z) & \text{if } z \in \overline{\mathbf{H}}, \\ g^*(z) & \text{if } z \in \mathbf{H}^* \end{cases} \tag{8.6.14}$$

defines a self-homeomorphism of $\overline{\mathbf{R}}^2$ which is K-quasiconformal in $\mathbf{H}$ and conformal in $\mathbf{H}^*$. Thus h is K-quasiconformal in $\overline{\mathbf{R}}^2$ and $D = h(\mathbf{H})$ is a K-quasidisk.

Finally, set

$$f^*(z) = h \circ r \circ h^{-1}(z) = g^* \circ r \circ f \circ g^{-1}(z) \tag{8.6.15}$$

for $z \in \overline{D}$ where $r(z) = \overline{z}$. Then $f^*(D) = D^*$, $f^*(z) = z$ for $z \in \partial D$, and

$$\frac{h_{D^*}(f^*(z_1), f^*(z_2))}{h_D(z_1, z_2)} = \frac{h_{\mathbf{H}}(f(w_1), f(w_2))}{h_{\mathbf{H}}(w_1, w_2)}$$

for $z_1, z_2 \in D$ where $w_j = g^{-1}(z_j) \in \mathbf{H}$. We conclude from (8.6.13) that f^* is a hyperbolic L-bilipschitz reflection in ∂D. □

8.7. Reflections and quasidisks

Theorem 8.6.7 completes a circle of implications to show the equivalence of five different characterizations of a quasidisk. We conclude this chapter by establishing three additional descriptions for this class of domains. The first two are concerned with reflections.

We will need the follwing lemma.

LEMMA 8.7.1. *Suppose that D and D' are simply connected domains in $\overline{\mathbf{R}}^2$ and that $f : D \to D'$ is a homeomorphism such that*

$$\frac{1}{L}\, h_D(z_1, z_2) \le h_{D'}(f(z_1), f(z_2)) \le L\, h_D(z_1, z_2) \tag{8.7.2}$$

for $z_1, z_2 \in D$ where $L < \infty$. Then f is an L^2-quasiconformal mapping.

PROOF. For $z_0 \in D \setminus \{\infty, f^{-1}(\infty)\}$ and $l > 1$ we can choose $\delta > 0$ so that

$$\frac{\rho_D(z_0)}{l} \le \frac{h_D(z, z_0)}{|z - z_0|} \le l\, \rho_D(z_0)$$

and

$$\frac{\rho_{D'}(f(z_0))}{l} \le \frac{h_{D'}(f(z), f(z_0))}{|f(z) - f(z_0)|} \le l\, \rho_{D'}(f(z_0))$$

when $0 < |z - z_0| < \delta$, where ρ_D and $\rho_{D'}$ denote the hyperbolic densities in D and D'. Then (8.7.2) implies that

$$\frac{|f(z) - f(z_0)|}{|z - z_0|} \le l^2\, \frac{h_{D'}(f(z), f(z_0))}{\rho_{D'}(f(z_0))}\, \frac{\rho_D(z_0)}{h_D(z, z_0)} \le L\, l^2 \frac{\rho_D(z_0)}{\rho_{D'}(f(z_0))},$$

$$\frac{|f(z) - f(z_0)|}{|z - z_0|} \ge \frac{1}{l^2}\, \frac{h_{D'}(f(z), f(z_0))}{\rho_{D'}(f(z_0))}\, \frac{\rho_D(z_0)}{h_D(z, z_0)} \ge \frac{1}{L\, l^2}\, \frac{\rho_D(z_0)}{\rho_{D'}(f(z_0))}$$

for $0 < |z - z_0| < \delta$ and we obtain

$$\begin{aligned} \frac{1}{L}\, \frac{\rho_D(z_0)}{\rho_{D'}(f(z_0))} &\le \liminf_{|z - z_0| \to 0} \frac{|f(z) - f(z_0)|}{|z - z_0|} \\ &\le \limsup_{|z - z_0| \to 0} \frac{|f(z) - f(z_0)|}{|z - z_0|} \le L\, \frac{\rho_D(z_0)}{\rho_{D'}(f(z_0))}. \end{aligned}$$

In particular,

$$H_f(z_0) = \limsup_{r \to 0} \frac{\sup_{|z - z_0| = r} |f(z) - f(z_0)|}{\inf_{|z - z_0| = r} |f(z) - f(z_0)|} \le L^2$$

and f is L^2-quasiconformal by Definition 1.1.3. □

This lemma together with Theorem 2.1.4 allows us to characterize the domains which admit a hyperbolic bilipschitz reflection.

THEOREM 2.1.11. *A domain D is a K-quasidisk if and only if it admits a hyperbolic L-bilipschitz reflection, where K and L depend only on each other.*

PROOF. Suppose that D is a K-quasidisk. Then the chain of implications established in Sections 8.1 through 8.6 and Theorem 8.6.7 imply that D admits a hyperbolic L-bilipschitz reflection where $L = L(K)$. Conversely if D admits a hyperbolic L-bilipschitz reflection, Theorem 2.1.4 and Lemma 8.7.1 imply that D is a K-quasidisk where $K = L$. □

The following result yields a Euclidean analogue of Theorem 2.1.11.

LEMMA 8.7.3. *Suppose that* $h : \overline{\mathbf{R}}^2 \to \overline{\mathbf{R}}^2$ *is* K*-quasiconformal with* $h(\infty) = \infty$ *and that*

$$\frac{1}{L} \le \frac{h_D(h(z_1), h(z_2))}{h_{D^*}(h(\overline{z}_1), h(\overline{z}_2))} \le L \tag{8.7.4}$$

for $z_1, z_2 \in \mathbf{H}$ *where* $D = h(\mathbf{H})$ *and* $D^* = h(\mathbf{H}^*)$. *Then*

$$\frac{1}{M} \le \frac{|h(z_1) - h(z_2)|}{|h(\overline{z}_1) - h(\overline{z}_2)|} \le M \tag{8.7.5}$$

for $z_1, z_2 \in \mathbf{H}$ *where* $M = M(K, L)$.

PROOF. Suppose that $z \in \mathbf{H}$. Then for each $z_0 \in \partial \mathbf{H}$,

$$|\overline{z} - z_0| = |z - z_0|$$

and

$$\frac{1}{c} \le \frac{|h(z) - h(z_0)|}{|h(\overline{z}) - h(z_0)|} \le c \tag{8.7.6}$$

by Theorem 1.3.4 where $c = e^{8K}$. In particular if we choose z_0 in (8.7.6) first so that

$$|h(z) - h(z_0)| = \operatorname{dist}(h(z), \partial D)$$

and then so that

$$|h(\overline{z}) - h(z_0)| = \operatorname{dist}(h(\overline{z}), \partial D^*),$$

we obtain

$$\frac{1}{c} \le \frac{\operatorname{dist}(h(z), \partial D)}{\operatorname{dist}(h(\overline{z}), \partial D^*)} \le c.$$

Thus

$$\frac{1}{4c} \le \frac{\rho_D(h(z))}{\rho_{D^*}(h(\overline{z}))} \le 4c \tag{8.7.7}$$

by (3.2.1) where ρ_D and ρ_{D^*} denote the hyperbolic densities in D and D^*. Since

$$\rho_D(h(z_0)) = \lim_{z \to z_0} \frac{h_D(h(z), h(z_0))}{|h(z) - h(z_0)|}$$

and

$$\rho_{D^*}(h(\overline{z}_0)) = \lim_{z \to z_0} \frac{h_{D^*}(h(\overline{z}), h(\overline{z}_0))}{|h(\overline{z}) - h(\overline{z}_0)|},$$

we conclude from (8.7.4) and (8.7.7) that

$$\frac{1}{M} < \liminf_{z \to z_0} \frac{|h(z) - h(z_0)|}{|h(\overline{z}) - h(\overline{z}_0)|} \le \limsup_{z \to z_0} \frac{|h(z) - h(z_0)|}{|h(\overline{z}) - h(\overline{z}_0)|} < M \tag{8.7.8}$$

for each $z_0 \in \mathbf{H}$ where $M > 4cL$. Inequality (8.7.6) shows that (8.7.8) also holds for $z_0 \in \overline{\mathbf{H}}$. Finally, since (8.7.8) is symmetric, it also holds for $z_0 \in \overline{\mathbf{H}}^*$ and hence for $z_0 \in \mathbf{R}^2$.

Suppose that γ is any arc in $\mathbf{R}^2$ with endpoints z_1, z_2. By (8.7.8) we can choose consecutive points $w_1, w_2, \ldots, w_{n+1} \in \gamma$ such that $w_1 = z_1$, $w_{n+1} = z_2$, and

$$\frac{1}{M} \le \frac{|h(w_j) - h(w_{j+1})|}{|h(\overline{w}_j) - h(\overline{w}_{j+1})|} \le M$$

for $j = 1, 2, \ldots, n$. If, in particular, $h(\gamma)$ is the segment joining $h(z_1)$ and $h(z_2)$, then

$$\begin{aligned} |h(\overline{z}_1) - h(\overline{z}_2)| &\leq \sum_{j=1}^{n} |h(\overline{w}_j) - h(\overline{w}_{j+1})| \leq M \sum_{j=1}^{n} |h(w_j) - h(w_{j+1})| \\ &= M|h(z_1) - h(z_2)|. \end{aligned}$$

Similarly

$$\begin{aligned} |h(z_1) - h(z_2)| &\leq \sum_{j=1}^{n} |h(w_j) - h(w_{j+1})| \leq M \sum_{j=1}^{n} |h(\overline{w}_j) - h(\overline{w}_{j+1})| \\ &= M|h(\overline{z}_1) - h(\overline{z}_2)| \end{aligned}$$

if $h(r(\gamma))$ is a segment where $r(z) = \overline{z}$. □

THEOREM 2.1.8 (Ahlfors [**5**]). *Suppose that $\infty \in \partial D$. Then D is a K-quasidisk if and only if it admits a Euclidean L-bilipschitz reflection, where K and L depend only on each other.*

PROOF. If f is a mapping which is L-bilipschitz with respect to the Euclidean metric, then f is L^2-quasiconformal by Definition 1.1.3 and D is a $K = K(L)$-quasidisk by Theorem 2.1.4.

Conversely if D is a K-quasidisk, then $D = h(\mathbf{H})$ where $h : \overline{\mathbf{R}}^2 \to \overline{\mathbf{R}}^2$ is K-quasiconformal and

$$f(z) = h \circ r \circ h^{-1}(z), \qquad r(z) = \overline{z},$$

is a hyperbolic L-bilipschitz reflection in ∂D where $L = L(K)$. If $w_1, w_2 \in D$ and if $z_j = h^{-1}(w_j)$, then $f(w_j) = h(\overline{z}_j)$ and we obtain

$$\frac{1}{M} \leq \frac{|f(w_1) - f(w_2)|}{|w_1 - w_2|} \leq M$$

from Lemma 8.7.3. □

8.8. Quasidisks and decomposability

Recall that a domain D is K'-quasiconformally decomposable if for each $z_1, z_2 \in D$ there exists a K'-quasidisk D' such that

$$z_1, z_2 \in D' \subset D. \tag{8.8.1}$$

We conclude this chapter by showing that quasiconformal decomposability characterizes the class of quasidisks.

THEOREM 2.5.3. *A simply connected domain D is a K-quasidisk if and only if it is K'-quasiconformally decomposable, where K and K' depend only on each other.*

PROOF. A K-quasidisk is clearly K'-quasiconformally decomposable with $K' = K$.

For the converse we may assume by means of a preliminary Möbius transformation that $D \subset \mathbf{R}^2$. Then for each $z_1, z_2 \in D$, there exists a K'-quasidisk and a hyperbolic segment γ joining z_1 and z_2 in $D' \subset D$ such that for each $z \in \gamma$,

$$\text{length}(\gamma) \leq c\,|z_1 - z_2|,$$
$$\min_{j=1,2} \text{length}(\gamma_j) \leq c\,\text{dist}(z, \partial D).$$

Here γ_1, γ_2 are the components of $\gamma \setminus \{z\}$ and $c = c(K')$. Thus D is a simply connected uniform domain and hence a K-quasidisk where K depends only on K'. $\square$

CHAPTER 9

Second series of implications

In the last chapter we proved seven ways of characterizing a quasidisk by means of a circle of implications. Two of the characterizations established there for a simply connected domain D are as follows.

D is a quasidisk if and only if it is a uniform domain.
D is a quasidisk if and only if it is linearly locally connected.

We now use this information to establish four additional characterizations by proving the following statements for a simply connected domain D.

1° If D is uniform, then $\sigma(D) > 0$.
2° If $\sigma(D) > 0$, then $\tau(D) > 0$.
3° If $\tau(D) > 0$, then D is linearly locally connected.
4° If D is uniform, then $L(D) > 1$.
5° If $L(D) > 1$, then D is linearly locally connected.
6° If D is uniform, then it has the min-max property.
7° If D has the min-max property, then D is linearly locally connected.

Implications 1°, 2°, and 3° concern domains D where a meromorphic function f is injective whenever its Schwarzian derivative S_f or pre-Schwarzian T_f is not large compared to the hyperbolic metric ρ_D. Implications 4° and 5° consider the same problem for functions f locally bilipschitz in D with small lipschitz constant. Finally, 6° and 7° involve a geometric property of the hyperbolic geodesics in D.

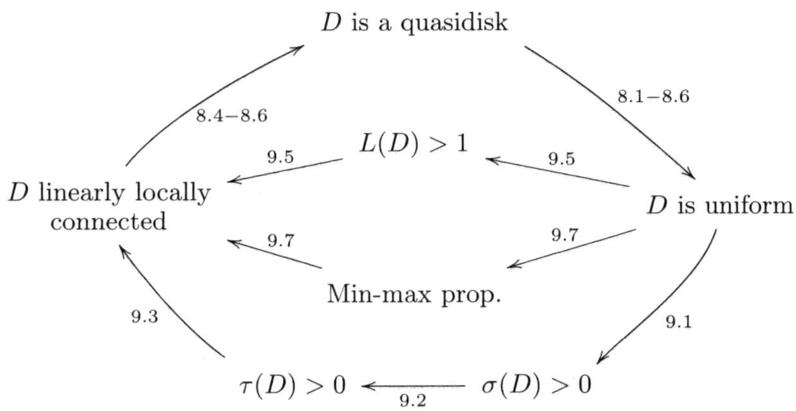

FIGURE 9.1

9.1. Uniform domains and Schwarzian derivatives

The arguments in this section are based in large part on the ideas of Martio and Sarvas in [**123**]. We begin with a lemma which compares the size of the pre-Schwarzian derivative $T_f = f''/f'$ with the values f assumes at two different points.

LEMMA 9.1.1. *Suppose that $z_1, z_2 \in D \subset \mathbf{R}^2$, that γ is a rectifiable open arc joining z_1, z_2 in D with midpoint z_0, and that $0 < c < 1$. If f is meromorphic and locally injective in D and if*

$$\left|\frac{f''(z)}{f'(z)}\right| \leq \frac{c}{\min(s, \operatorname{length}(\gamma) - s)} \tag{9.1.2}$$

for $z \in \gamma$, where s is the arclength of γ from z_1 to z, then

$$\left|\frac{f(z_1) - f(z_2)}{f'(z_0)} - (z_1 - z_2)\right| \leq \frac{c}{1-c} \operatorname{length}(\gamma).$$

PROOF. By the triangle inequality it is sufficient to prove that

$$\left|\frac{f(z_j) - f(z_0)}{f'(z_0)} - (z_j - z_0)\right| \leq \frac{1}{2}\frac{c}{1-c} \operatorname{length}(\gamma) \tag{9.1.3}$$

for $j = 1, 2$. By symmetry we need only consider the case where $j = 1$.

Now (9.1.2) implies that f''/f' and hence f are finite at each $z \in \gamma$. For $z \in \gamma$ let

$$g(z) = \int_{z_0}^{z} \frac{f''}{f'}(\zeta)\, d\zeta,$$

where the integral is taken along γ. Then

$$e^{g(z)} = \frac{f'(z)}{f'(z_0)}$$

and

$$\frac{f(z) - f(z_0)}{f'(z_0)} - (z - z_0) = \int_{z_0}^{z} \left(e^{g(\zeta)} - 1\right) d\zeta$$

for $z \in \gamma$. If $z \in \gamma$ is between z_1 and z_0, then

$$\left|\frac{f''}{f'}(z)\right| \leq \frac{c}{s},$$

by (9.1.2), whence

$$|g(z)| \leq \int_{z_0}^{z} \left|\frac{f''}{f'}(\zeta)\right| |d\zeta| \leq \int_{s}^{a} \frac{c}{\sigma}\, d\sigma = c \log\left(\frac{a}{s}\right),$$

where for convenience of notation we let $a = \operatorname{length}(\gamma)/2$. Therefore

$$\left|e^{g(z)} - 1\right| \leq e^{|g(z)|} - 1 \leq \left(\frac{a}{s}\right)^{c} - 1$$

and we obtain

$$\begin{aligned}
\left|\frac{f(z) - f(z_0)}{f'(z_0)} - (z - z_0)\right| &\leq \int_{z_0}^{z} \left|e^{g(\zeta)} - 1\right| |d\zeta| \\
&\leq \int_{0}^{a} \left(\left(\frac{a}{\sigma}\right)^{c} - 1\right) d\sigma \\
&= \frac{1}{2}\frac{c}{1-c} \operatorname{length}(\gamma)
\end{aligned}$$

for $z \in \gamma$ between z_1 and z_0. This inequality then implies that f is bounded near z_1 and hence analytic at z_1. Thus we can let $z \to z_1$ along γ to get (9.1.3). $\square$

The following result allows us to replace the pre-Schwarzian derivative $T_f = f''/f'$ in Lemma 9.1.1 with the Schwarzian derivative S_f.

LEMMA 9.1.4. *If u and v are absolutely continuous in each closed subinterval of $[a,b)$, if $u(a) = 0$, and if*

$$u' < uv$$

a.e. in $[a,b)$, then $u \le 0$ in $[a,b)$.

PROOF. Let

$$w(t) = u(t)\exp\left(-\int_a^t v(s)\,ds\right).$$

Then w is absolutely continuous in each closed subinterval of $[a,b)$,

$$w'(t) = \exp\left(-\int_a^t v(s)\,ds\right)(u'(t) - u(t)v(t)) \le 0$$

a.e. in $[a,b)$, and hence

$$w(t) \le w(a) = 0, \qquad u(t) = w(t)\exp\left(\int_a^t v(s)\,ds\right) \le 0$$

in $[a,b)$. $\square$

We then have the following analogue of Lemma 9.1.1 for the Schwarzian derivative S_f.

LEMMA 9.1.5. *Suppose that $z_1, z_2 \in D$, that γ is a rectifiable open arc joining z_1, z_2 in D with midpoint z_0, and that $0 < c < \frac{1}{2}$. If f is meromorphic and locally injective in D with $f''(z_0) = 0$ and if*

$$|S_f(z)| \le \frac{c}{\min(s, \text{length}(\gamma) - s)^2}, \tag{9.1.6}$$

for $z \in \gamma$, where s is the arclength of γ from z_1 to z, then

$$\left|\frac{f(z_1) - f(z_2)}{f'(z_0)} - (z_1 - z_2)\right| \le \frac{2c}{1-2c}\,\text{length}(\gamma).$$

PROOF. By Lemma 9.1.1 it is sufficient to show that

$$\left|\frac{f''(z)}{f'(z)}\right| \le \frac{2c}{\min(s, \text{length}(\gamma) - s)} \tag{9.1.7}$$

for each $z \in \gamma$. Again by symmetry we need only prove (9.1.7) for the case where z lies between z_0 and z_2, i.e., for $s \in [a, \text{length}(\gamma))$ where again we let $a = \text{length}(\gamma)/2$.

Since $f''(z_0) = 0$, f is finite at z_0 and there exists $t \in (a, \text{length}(\gamma))$ such that f is finite at $z(s)$ for $s \in [a,t)$, where $z(s)$ is the arclength representation of γ. Let b denote the supremum of all such numbers t, and for $s \in [a,b)$ let

$$\phi(s) = \frac{2c}{\text{length}(\gamma) - s}, \qquad \psi(s) = \left|\frac{f''}{f'}(z(s))\right| + \phi(a).$$

Then ϕ and ψ are absolutely continuous in each closed subinterval of $[a,b)$ with

$$\phi'(s) - \frac{1}{2}\,\phi(s)^2 = \frac{2c(1-c)}{(\text{length}(\gamma) - s)^2} \ge \frac{c}{(\text{length}(\gamma) - s)^2}$$

and

$$\begin{aligned}\psi'(s) - \frac{1}{2}\psi(s)^2 &< \left| \left(\frac{f''}{f'}(z(s)) \right)' z'(s) \right| - \frac{1}{2} \left| \frac{f''}{f'}(z(s)) \right|^2 \\ &\leq \left| \left(\frac{f''}{f'}(z(s)) \right)' - \frac{1}{2} \left(\frac{f''}{f'}(z(s)) \right)^2 \right| = |S_f(z(s))| \\ &\leq \frac{c}{(\text{length}(\gamma) - s)^2}\end{aligned}$$

a.e. in $[a, b)$. Thus

$$\psi'(s) - \phi'(s) < \frac{1}{2}(\psi(s)^2 - \phi(s)^2) = (\psi(s) - \phi(s)) \left(\frac{\psi(s) + \phi(s)}{2} \right)$$

a.e. in $[a, b)$, and we can apply Lemma 9.1.4 with $u = \psi - \phi$ and $v = \frac{1}{2}(\psi + \phi)$ to conclude that

$$\left| \frac{f''}{f'}(z(s)) \right| < \psi(s) \leq \phi(s) = \frac{2c}{\min(s, \text{length}(\gamma))} \tag{9.1.8}$$

for $s \in [a, b)$.

Finally, we claim that $b = \text{length}(\gamma)$. Otherwise, (9.1.8) would imply that f''/f' is bounded near $z(b)$ and hence that f is analytic at $z(b)$. In particular, we would then get $b' \in (b, \text{length}(\gamma))$ such that f is analytic at $z(s)$ for $s \in [a, b')$, contradicting the way b was chosen. This completes the proof of (9.1.7) and hence of Lemma 9.1.5. □

We show now that if D is uniform, then $\sigma(D) > 0$, i.e., functions analytic in D are injective if their Schwarzian derivatives are small compared with the square of the hyperbolic metric in D.

THEOREM 9.1.9. *Suppose D is a simply connected domain in $\mathbf{R}^2$ and suppose there exists a constant $c > 1$ such that each $z_1, z_2 \in D$ can be joined by an arc $\gamma \subset D$ so that*

$$\text{length}(\gamma) \leq c\,|z_1 - z_2|, \tag{9.1.10}$$

$$\min_{j=1,2} \text{length}(\gamma_j) \leq c\,\text{dist}(z, \partial D) \tag{9.1.11}$$

for each $z \in \gamma$, where γ_1, γ_2 are the components of $\gamma \setminus \{z\}$. If f is meromorphic and locally injective in D and if

$$\sup_{z \in D} |S_f(z)| \rho_D(z)^{-2} < \frac{1}{16\, c^3}, \tag{9.1.12}$$

then f is injective in D. Hence, in particular,

$$\sigma(D) \geq \frac{1}{16\, c^3}. \tag{9.1.13}$$

PROOF. Choose $z_1, z_2 \in D$ and let γ be an arc joining these points for which (9.1.10) and (9.1.11) hold. Then by (3.2.1) and (9.1.13),

$$\begin{aligned}|S_f(z)| &\leq \frac{1}{16\, c^3}\, \rho_D(z)^2 \leq \frac{1}{4\, c^3}\, \frac{1}{\text{dist}(z, \partial D)^2} \\ &\leq \frac{1}{4\, c}\, \frac{1}{\min(s, \text{length}(\gamma) - s)^2}\end{aligned} \tag{9.1.14}$$

for $z \in \gamma$, where s is the arclength of γ measured from z_1 to z.

Let z_0 be the midpoint of γ. If $f''(z_0) = 0$, then

$$\begin{aligned}\left|\frac{f(z_1) - f(z_2)}{f'(z_0)} - (z_1 - z_2)\right| &\le \frac{1}{2\,c - 1}\,\text{length}(\gamma)\\ &\le \frac{c}{2\,c-1}\,|z_1 - z_2| < |z_1 - z_2|\end{aligned}$$

by (9.1.14) and Lemma 9.1.5 with $1/4c$ in place of c and by (9.1.10). Hence $f(z_1) \neq f(z_2)$.

If $f''(z_0) \neq 0$, we can choose a Möbius transformation g so that $h''(z_0) = 0$ where $h = g \circ f$. Then $S_h = S_f$ and we can apply the above argument to h to conclude that $h(z_1) \neq h(z_2)$, whence $f(z_1) \neq f(z_2)$. Thus f is injective in D. □

9.2. Schwarzian and pre-Schwarzian derivatives

Suppose that D is a simply connected domain and that f is any function which is analytic and locally injective in D. We show here that if there is a constant $a > 0$ so that f is injective whenever

$$|S_f(z)| \le a\,\rho_D(z)^2$$

in D, then there is a second constant $b > 0$ so that f is injective whenever

$$|T_f(z)| \le b\,\rho_D(z)$$

in D. This fact is an easy consquence of the following result. Cf. Duren-Shapiro-Shields [**35**], Wirths [**167**].

THEOREM 9.2.1. *Suppose that f is analytic in a simply connected domain $D \subset \mathbf{R}^2$. If*

$$\sup_D |f(z)|\,\rho_D(z)^{-1} \le c, \tag{9.2.2}$$

then

$$\sup_D |f'(z)|\,\rho_D(z)^{-2} \le 5c. \tag{9.2.3}$$

PROOF. Suppose first that $D = \mathbf{B}$, fix $z \in \mathbf{B}$, and choose $r > 0$ so that $2\,r^2 = 1 + |z|^2$, whence

$$1 - r^2 = r^2 - |z|^2 = \rho_{\mathbf{B}}(z)^{-1}.$$

Then

$$|f(\zeta)| \le c\,\frac{2}{1 - |\zeta|^2}$$

for $\zeta \in \mathbf{B}$ and with the Cauchy and Poisson integral formulas we obtain

$$\begin{aligned}|f'(z)| &= \left|\frac{1}{2\,\pi i}\int_{|\zeta|=r}\frac{f(\zeta)}{(\zeta - z)^2}\,d\zeta\right|\\ &\le \frac{2c}{1-r^2}\,\frac{r}{r^2 - |z|^2}\,\frac{1}{2\pi}\int_0^{2\pi}\frac{r^2 - |z|^2}{|r\,e^{i\,\theta} - z|^2}\,d\theta\\ &= \frac{2c}{1-r^2}\,\frac{r}{r^2 - |z|^2} < 2c\,\rho_{\mathbf{B}}(z)^2.\end{aligned}$$

For the general case set $g(z) = f(\phi(z))\phi'(z)$ where $\phi : \mathbf{B} \to D$ is conformal. Then g is analytic in $\mathbf{B}$,

$$|g(z)| \le c\,\rho_D(\phi(z))|\phi'(z)| = c\,\rho_{\mathbf{B}}(z)$$

for $z \in \mathbf{B}$, and

$$\begin{aligned} |f'(\phi(z))\,\phi'(z)^2 + f(\phi(z))\phi''(z)| &= |g'(z)| \le 2c\,\rho_{\mathbf{B}}(z)^2 \\ &= 2c\,\rho_D(\phi(z))^2\,|\phi'(z)|^2 \end{aligned}$$

by what was proved above. Hence

$$\begin{aligned} |f'(\phi(z))| &\le |f(\phi(z))| \left|\frac{\phi''(z)}{\phi'(z)^2}\right| + 2c\,\rho_D(\phi(z))^2 \\ &\le c\,\rho_D(\phi(z)) \left|\frac{\phi''(z)}{\phi'(z)^2}\right| + 2c\,\rho_D(\phi(z))^2 \le 5c\,\rho_D(\phi(z))^2, \end{aligned}$$

since from a well-known distortion theorem for conformal maps of $\mathbf{B}$ we obtain

$$\left| z\,\frac{\phi''(z)}{\phi'(z)} - \frac{2|z|^2}{1-|z|^2} \right| \le \frac{4|z|}{1-|z|^2},$$

whence

$$\left|\frac{\phi''(z)}{\phi'(z)}\right| \le \frac{2|z|+4}{1-|z|^2} \le 3\,\rho_{\mathbf{B}}(z) = 3\,\rho_D(\phi(z))|\phi'(z)|.$$

See, for example, Duren [**33**], Hayman [**78**]. □

Corollary 9.2.4. *Suppose that f is analytic and locally injective in a simply connected domain $D \subset \mathbf{R}^2$. If*

$$\sup_{z\in D} |T_f(z)|\,\rho_D(z)^{-1} \le c,$$

then

$$\sup_{z\in D} |S_f(z)|\,\rho_D(z)^{-2} \le 5c + \tfrac{1}{2}\,c^2.$$

Proof. By hypothesis, T_f is analytic with

$$S_f(z) = T_f'(z) - \tfrac{1}{2}T_f(z)^2$$

in D. Hence

$$|S_f(z)|\,\rho_D(z)^{-2} \le |T_f'(z)|\rho_D(z)^{-2} + \tfrac{1}{2}\left(|T_f(z)|\,\rho_D(z)^{-1}\right)^2 \le 5c + \tfrac{1}{2}c^2$$

by Theorem 9.2.1. □

The following result is an immediate consequence of Corollary 9.2.4.

Corollary 9.2.5. *If $D \subset \mathbf{R}^2$ is simply connected with $\sigma(D) > 0$, then*

$$\tau(D) \ge \sqrt{25 + \sigma(D)} - 5 > 0.$$

9.3. Pre-Schwarzian derivatives and local connectivity

We saw in the previous section that the pre-Schwarzian radius of injectivity $\tau(D)$ of a simply connected domain D is positive whenever its Schwarzian radius of injectivity $\sigma(D)$ is positive. We show here that the domain D is linearly locally connected whenever $\tau(D) > 0$.

The proof of this fact depends on the following four geometric lemmas.

LEMMA 9.3.1 (Gehring-Martio [**64**]). *Suppose that $c > 1$ and that there exist two points in $D \cap \overline{\mathbf{B}}(z_0, r)$ which cannot be joined in $D \cap \overline{\mathbf{B}}(z_0, cr)$. Then there exist points $z_1, z_2 \in D$ and $w_0 \in \partial\mathbf{B}(z_0, cr) \setminus D$ such that*

$$|h(z_1) - h(z_2) - 2\pi i| \le \frac{2}{c-1}, \tag{9.3.2}$$

where $h(z) = \log(z - w_0)$.

PROOF. Let z_1', z_2' be two points in $D \cap \overline{\mathbf{B}}(z_0, r)$ which cannot be joined in $D \cap \overline{\mathbf{B}}(z_0, cr)$, and let α' be the segment and β' a rectifiable arc which join z_1', z_2' in $\mathbf{R}^2$ and D, respectively. We may choose β' so that it intersects α' in a finite set of points E. Then there exist two adjacent points $z_1, z_2 \in E$ which cannot be joined in $D \cap \overline{\mathbf{B}}(z_0, cr)$. Let α and β denote the parts of α' and β' between z_1 and z_2. Then $\gamma = \alpha \cup \beta$ is a Jordan curve and we denote by D_0 the bounded component of $\overline{\mathbf{R}}^2 \setminus \gamma$. The fact that z_1, z_2 cannot be joined in $D \cap \overline{\mathbf{B}}(z_0, cr)$ and a simple topological argument based on Kerékjártó's theorem (Newman [**140**]) imply the existence of a point w_0 such that

$$w_0 \in (\overline{\mathbf{R}}^2 \setminus D) \cap \partial\mathbf{B}(z_0, cr) \cap D_0.$$

For the details see Gehring [**49**]. Since D is simply connected, we can choose an analytic branch of $h(z) = \log(z - w_0)$ in D, and

$$h(z_1) - h(z_2) = \int_\beta h'(z)\, dz = 2\pi i\, n(\gamma, w_0) - \int_\alpha \frac{dz}{z - w_0},$$

where $n(\gamma, w_0)$ is the winding number of γ with respect to w_0. Now

$$n(\gamma, w_0) = n = \pm 1,$$

and hence

$$|h(z_1) - h(z_2) - 2\pi n i| \le \int_\alpha \frac{|dz|}{|z - w_0|}. \tag{9.3.3}$$

Since $\alpha \subset \overline{\mathbf{B}}(z_0, r)$ and $w_0 \in \partial\mathbf{B}(z_0, cr)$,

$$\int_\alpha \frac{|dz|}{|z - w_0|} \le \frac{\text{length}(\alpha)}{(c-1)r} \le \frac{2}{c-1}$$

and (9.3.2) follows from (9.3.3) if $n = 1$. If $n = -1$, the result follows by interchanging z_1 and z_2. □

LEMMA 9.3.4 (Astala-Gehring [**14**]). *Suppose that $c > 2$, that $z_0 \in \mathbf{R}^2$, and that $0 < r < \infty$. If there are points in $D \cap \overline{\mathbf{B}}(z_0, r)$ which cannot be joined in $D \cap \overline{\mathbf{B}}(z_0, cr)$, then*

$$\tau(D) \le \frac{2}{\pi(c-1) - 1}. \tag{9.3.5}$$

PROOF. By Lemma 9.3.1 there exist points $z_1, z_2 \in D$ and $w_0 \in \mathbf{R}^2 \setminus D$ such that

$$|h(z_1) - h(z_2) - 2\pi i| \le \frac{2}{c-1}, \tag{9.3.6}$$

where h is any continuous branch of $\log(z - w_0)$ in D. Set

$$f(z) = e^{a\,h(z)}, \qquad a = \frac{2\pi i}{h(z_1) - h(z_2)}.$$

Then f is analytic with $f' \neq 0$ in D and

$$\left|\frac{f''}{f'}\right| = \frac{|a-1|}{|z - w_0|} \le 2\,|a-1|\,\rho_D \le \frac{2}{\pi(c-1)-1}\,\rho_D \tag{9.3.7}$$

by the Koebe distortion theorem and (9.3.6). Since

$$\frac{f(z_1)}{f(z_2)} = e^{a(h(z_1)-h(z_2))} = 1,$$

f is not injective in D and (9.3.5) follows from the definition of $\tau(D)$. □

LEMMA 9.3.8 (Astala-Gehring [**14**]). *Suppose that* $h : \mathbf{B} \to D$ *is conformal with* $h(0) = 0$ *and that*

$$\operatorname{dist}(h(0), \partial D) = 1.$$

Suppose also that $-1 < x_1 < 0 < x_2 < 1$ *and that*

$$|h(x)| \le |h(x_1)| = |h(x_2)| = b > 0 \tag{9.3.9}$$

for $x_1 \le x \le x_2$. *Then*

$$\log b < 3\left|\int_{x_1}^{x_2} \frac{h'(x)}{h(x)}\,x\,dx\right|. \tag{9.3.10}$$

PROOF. We may assume that $b > 1$ since otherwise (9.3.10) follows trivially. Choose y_1 and y_2 so that $x_1 < y_1 < 0 < y_2 < x_2$ and

$$|h(x)| \le |h(y_1)| = |h(y_2)| = 1 \tag{9.3.11}$$

for $y_1 \le x \le y_2$. Then $zh'(z)/h(z)$ is analytic in $\mathbf{B}$ and

$$\frac{d}{dx}\log|h(x)| = \operatorname{Re}\left(\frac{h'(x)}{h(x)}\right)$$

for $-1 < x < 1$, $x \neq 0$. Thus

$$\begin{aligned}\left|\int_{x_1}^{x_2} \frac{h'(x)}{h(x)}\,x\,dx\right| &\ge \int_{x_1}^{x_2} \operatorname{Re}\left(\frac{h'(x)}{h(x)}\right) x\,dx \\ &= (x_2 - x_1)\log b + \int_{x_1}^{x_2} \log\frac{1}{|h(x)|}\,dx \\ &\ge (y_2 - y_1)\log b + \left(\int_{x_1}^{y_1} + \int_{y_2}^{x_2}\right)\log\frac{b}{|h(x)|}\,dx \\ &\ge (y_2 - y_1)\log b\end{aligned} \tag{9.3.12}$$

by integration by parts, (9.3.9) and (9.3.11). Since

$$|h'(0)| \le 4\operatorname{dist}(h(0), \partial D) = 4,$$

standard distortion theorems applied to $h(z)/h'(0)$ imply that

$$1 = |h(y_j)| \le 4 \, \frac{|y_j|}{(1-|y_j|)^2} \qquad \text{and hence} \qquad |y_j| > \frac{1}{6}$$

for $j = 1, 2$. This, together with (9.3.12), yields inequality (9.3.10). □

LEMMA 9.3.13 (Astala-Gehring [**14**]). *Suppose that $c > 2e^{3\pi} + 1$, that $w_0 \in \mathbf{R}^2$, and that $0 < r < \infty$. If there are points in $D \setminus \mathbf{B}(w_0, cr)$ which cannot be joined in $D \setminus \mathbf{B}(w_0, r)$, then*

$$\tau(D) \le 15\,\pi \left(\log \frac{c-1}{2}\right)^{-1}. \tag{9.3.14}$$

PROOF. Suppose that w_1, w_2 are points in $D \setminus \mathbf{B}(w_0, cr)$ which cannot be joined in $D \setminus \mathbf{B}(w_0, r)$ and let β denote the hyperbolic geodesic joining w_1 and w_2 in D. The hypotheses imply that

$$\beta \cap \mathbf{B}(w_0, r) \ne \emptyset \qquad \text{and} \qquad \overline{\mathbf{B}}(w_0, r) \setminus D \ne \emptyset.$$

Thus we can choose $z_0 \in \beta$ such that

$$\operatorname{dist}(z_0, \partial D) \le |z_0 - w_0| + \operatorname{dist}(w_0, \partial D) < 2r$$

where

$$|w_j - z_0| \ge |w_j - w_0| - |z_0 - w_0| > \frac{c-1}{2} \operatorname{dist}(z_0, \partial D) = d$$

for $j = 1, 2$.

Let α denote the component of $\beta \cap \mathbf{B}(z_0, d)$ which contains z_0 and let z_1, z_2 denote the endpoints of α. Then

$$|z - z_0| \le |z_1 - z_0| = |z_2 - z_0| = \frac{c-1}{2} \operatorname{dist}(z_0, \partial D) \tag{9.3.15}$$

for $z \in \alpha$.

If D' denotes the image of D under the similarity mapping

$$w = \frac{(z - z_0)}{\operatorname{dist}(z_0, \partial D)},$$

then it is easy to check that $\tau(D) = \tau(D')$. Hence we may assume without loss of generality that $z_0 = 0$ and $\operatorname{dist}(z_0, \partial D) = 1$. Let g map D conformally onto $\mathbf{B}$ so that $g(0) = 0$ and $0 < g(z_2) = x_2 < 1$. Then $-1 < g(z_1) = x_1 < 0$. Since $\mathbf{B} \subset D$, $g(\mathbf{B}) \subset \mathbf{B}$ and the Schwarz lemma implies that

$$\left|\frac{g(z)}{z}\right| \le \frac{\min(|z|, 1)}{|z|} \le \frac{2}{|z| + 1} \le \frac{2}{\operatorname{dist}(z, \partial D)} \le 4\rho_D(z) \tag{9.3.16}$$

and

$$\frac{|g'(z)|}{1 - |g(z)|} \le \rho_D(z) \tag{9.3.17}$$

in D.

Now suppose $\tau(D) > 0$ and choose $a > 0$ so that

$$a < \min(1, \frac{\tau(D)}{5}), \tag{9.3.18}$$

and for each $w \in \mathbf{B}$ let

$$f(z) = z\, e^{aw\, G(z)}, \qquad G(z) = \int_0^z \frac{g(\zeta)}{\zeta}\, d\zeta.$$

Then f is analytic in D with $f' \neq 0$ and

$$\left|\frac{f''}{f'}\right| = a\,|w|\left|\frac{g(z)}{z} + \frac{g'(z)}{1+aw\,g(z)}\right|$$
$$\leq a\left(\left|\frac{g(z)}{z}\right| + \frac{|g'(z)|}{1-|g(z)|}\right) \leq 5\,a\,\rho_D(z)$$

in D. Since $5\,a < \tau(D)$, f is injective and

$$1 \neq \frac{f(z_2)}{f(z_1)} = \frac{z_2}{z_1}\,e^{aw\,I} = e^{i\theta + aw\,I}$$

for all $w \in \mathbf{B}$ where $|\theta| \leq \pi$,

$$I = \int_{z_1}^{z_2} \frac{g(z)}{z}\,dz = \int_{x_1}^{x_2} \frac{h'(x)}{h(x)}\,x\,dx,$$

and $h = g^{-1}$. Thus $-i\theta/aI \notin \mathbf{B}$, whence

$$a \leq \frac{\pi}{|I|} < 3\pi\left(\log\frac{c-1}{2}\right)^{-1}$$

by Lemma 9.3.8 with $b = (c-1)/2$. Taking the supremum over all a which satisfy (9.3.18) yields

$$\min(1, \frac{\tau(D)}{5}) \leq 3\pi\left(\log\frac{c-1}{2}\right)^{-1}$$

by our choice of c, and we obtain (9.3.14). □

THEOREM 9.3.19. *Suppose $D \subset \mathbf{R}^2$ is simply connected and $\tau(D) > 0$. Then D is c-linearly locally connected where $c = c(\tau(D))$.*

PROOF. By hypothesis, a function f is injective in D if f is analytic and locally injective in D with

$$\sup_{z \in D}\left|\frac{f''(z)}{f'(z)}\right|\rho_D(z)^{-1} < \tau(D).$$

We must show that this hypothesis implies there exists a constant $c = c(\tau(D)) > 1$ such that D is c-locally connected, that is, for each $z_0 \in D$ and each $r > 0$

$$D \cap \mathbf{B}(z_0, r) \text{ lies in a component of } D \cap \mathbf{B}(z_0, cr), \tag{9.3.20}$$

$$D \setminus \overline{\mathbf{B}}(z_0, r) \text{ lies in a component of } D \setminus \overline{\mathbf{B}}(z_0, r/c). \tag{9.3.21}$$

Lemma 9.3.4 implies that points in $D \cap \mathbf{B}(z_0, r)$ can be joined in $D \cap \mathbf{B}(z_0, cr)$ and hence that (9.3.20) holds provided that

$$c > \frac{\tau(D)+2}{\pi\tau(D)} + 1.$$

Lemma 9.3.13 implies that points in $D \setminus \overline{\mathbf{B}}(z_0, r)$ can be joined in $D \setminus \overline{\mathbf{B}}(z_0, r/c)$ and hence that (9.3.21) holds provided that

$$c > 2\,\exp\left(\frac{15\pi}{\tau(D)}\right) + 1.$$

This completes the proof. □

9.4. Uniform domains are rigid

Suppose that $D \subset \mathbf{R}^2$ is a simply connected domain. In the last three sections we showed that D is a quasidisk if and only if each function f, analytic and locally injective in D, is globally injective whenever its Schwarzian derivative S_f or pre-Schwarzian derivative T_f is small relative to the hyperbolic density ρ_D, that is, if $\sigma(D) > 0$ or $\tau(D) > 0$.

We shall establish next the analogue of these results for bilipschitz maps. We begin with some preliminary results due to F. John and R. Nevanlinna.

LEMMA 9.4.1 (John [**90**]). *Suppose that $f : \overline{\mathbf{B}}(z_0, r) \to \mathbf{R}^2$ is L-bilipschitz. Then there exists an angle θ such that*

$$|f(z) - g(z)| \leq a(L-1)r \tag{9.4.2}$$

for $z \in \overline{\mathbf{B}}(z_0, r)$ where $a < 20$,

$$g(z) = E(z - z_0) + f(z_0),$$

and $E(z) = e^{i\theta} z$ or $e^{i\theta}\bar{z}$.

PROOF. By means of a preliminary translation and scaling we may assume that $r = 1$ and $f(z_0) = z_0 = 0$. We may also assume that

$$L < L_0 = \frac{2}{\sqrt{3}} \tag{9.4.3}$$

since otherwise

$$|f(z) - z| \leq L + 1 \leq (1 + \frac{2}{L_0 - 1})(L-1) \leq a(L-1)$$

and (9.4.2) will follow trivially with $E(z) = z$.

Before defining the map E when $L < L_0$, we establish an inequality that will be used repeatedly in what follows. Since f is L-bilipschitz with $f(0) = 0$,

$$-(L^2-1)|z_j|^2 \leq |f(z_j)|^2 - |z_j|^2 \leq (L^2-1)|z_j|^2$$

for $j = 1, 2$,

$$-(L^2-1)|z_1 - z_2|^2 \leq |f(z_1) - f(z_2)|^2 - |z_1 - z_2|^2 \leq (L^2-1)|z_1 - z_2|^2,$$

and by adding these inequalities we obtain

$$2\,|\mathrm{Re}(f(z_1)\overline{f(z_2)} - z_1\bar{z}_2)| \leq (L^2-1)(|z_1 - z_2|^2 + |z_1|^2 + |z_2|^2) \tag{9.4.4}$$

for $z_1, z_2 \in \mathbf{B}$. Next let $F(z) = F(x + iy) = u + iv$ be the linear transformation $F : \mathbf{R}^2 \to \mathbf{R}^2$ where

$$u = \mathrm{Re}\, f(1)\, x + \mathrm{Re}\, f(i)\, y,$$
$$v = \mathrm{Im}\, f(1)\, x + \mathrm{Im}\, f(i)\, y,$$

and let F also denote the associated matrix of F. We may assume that $\det(F) \geq 0$ by replacing $f(z)$ by $\overline{f(z)}$ if necessary.

We will estimate the semiaxes of the ellipse $F(\mathbf{B})$ by looking at the eigenvalues of $F^T F - I$ from which bounds are easily obtained. Here

$$F^T F = \begin{pmatrix} |f(1)|^2 & \mathrm{Re}(f(1)\overline{f(i)}) \\ \mathrm{Re}(f(1)\overline{f(i)}) & |f(i)|^2 \end{pmatrix}$$

and by (9.4.4)

$$|(F^T F - I)z|^2 = |(|f(1)|^2 - 1)\,x + \mathrm{Re}(f(1)\overline{f(i)})\,y|^2 \\ + |\mathrm{Re}(f(1)\overline{f(i)})\,x|^2 + |(|f(i)|^2 - 1)\,y|^2 \le 9(L^2-1)^2|z|^2,$$

whence

$$|(F^T F - I)\,z| \le 3(L^2-1)|z| < 3(L_0^2 - 1)|z| = |z| \tag{9.4.5}$$

by (9.4.3). Thus the eigenvalues of $F^T F - I$ lie in $(-1,1)$ and the eigenvalues of $F^T F$ are positive. In particular, $F(\mathbf{B})$ is an ellipse with semiaxes λ_1, λ_2, where

$$\sqrt{1 - 3(L^2-1)} \le \lambda_1 \le \lambda_2 \le \sqrt{1 + 3(L^2-1)}, \tag{9.4.6}$$

and there exist angles α, β such that

$$F(e^{i\,\alpha}) = \lambda_1\, e^{i\,\beta}, \qquad F(i\,e^{i\,\alpha}) = i\,\lambda_2\, e^{i\,\beta}.$$

More specifically, $F(z) = T(e^{i\,\theta} z)$ where $\theta = \beta - \alpha$ and $T : \mathbf{R}^2 \to \mathbf{R}^2$ is the linear transformation with

$$T(e^{i\,\beta}) = \lambda_1\, e^{i\,\beta}, \qquad T(i\,e^{i\,\beta}) = i\,\lambda_2\, e^{i\,\beta}.$$

Here

$$|T(z) - z| = |(T - I)z| \le b|z| \tag{9.4.7}$$

where

$$b = 1 - \sqrt{1 - 3(L^2-1)} \le 3(L^2-1). \tag{9.4.8}$$

Finally, since

$$f(1) = F(1) = T(e^{i\theta}), \qquad f(i) = F(i) = T(i\,e^{i\,\theta}),$$

(9.4.7) implies that

$$|f(1) - e^{i\,\theta}| \le b, \qquad |f(i) - i\,e^{i\,\theta}| \le b. \tag{9.4.9}$$

Then for $|z| \le 1$ we obtain

$$\begin{aligned} |f(z) - e^{i\,\theta}\,z|^2 &= |f(z)\,e^{-i\,\theta} - z|^2 \\ &= |\mathrm{Re}(f(z)e^{-i\,\theta} - z)|^2 + |\mathrm{Im}(f(z)e^{-i\,\theta} - z)|^2 \\ &= |\mathrm{Re}(f(z)e^{-i\,\theta} - z)|^2 + |\mathrm{Re}(i\,f(z)e^{-i\,\theta} - i\,z)|^2. \end{aligned}$$

Next the fact that $|f(z)| \le L$, (9.4.9), (9.4.4), and (9.4.8) imply that

$$\begin{aligned} |\mathrm{Re}(f(z)(e^{-i\,\theta} - z))| &\le |f(z)|\,|e^{-i\,\theta} - \overline{f(1)}| + |\mathrm{Re}(f(z)\overline{f(1)} - z)| \\ &\le Lb + (L^2-1)(1 + |z|^2 + |z-1|^2)/2 \\ &\le 3(L+1)(L^2-1) \end{aligned}$$

and similarly

$$\begin{aligned} |\mathrm{Re}(i\,f(z)e^{-i\,\theta} - i\,z)| &\le |f(z)|\,|i\,e^{-i\,\theta} + \overline{f(i)}| + |\mathrm{Re}(f(z)\overline{f(i)} + iz)| \\ &\le Lb + (L^2-1)(1 + |z|^2 + |z-i|^2)/2 \\ &\le 3(L+1)(L^2-1). \end{aligned}$$

We conclude that

$$|f(z) - e^{i\,\theta}\,z| \le 3\sqrt{2}(L+1)^2(L-1) < a(L-1),$$

where $a < 20$ from (9.4.3). □

LEMMA 9.4.10. *If E is a convex set in $\mathbf{R}^2$ and if $f : E \to \mathbf{R}^2$ is locally L-lipschitz in E, then f is L-lipschitz in E.*

PROOF. Suppose that $z_0, w_0 \in E$. Then $[z_0, w_0]$ is a compact subset of E and there exists a $\delta > 0$ such that

$$|f(z) - f(w)| \leq L|z - w| \tag{9.4.11}$$

whenever $z, w \in E$ and $|z - w| < \delta$. Next choose an ordered sequence of points $z_1, z_2, \ldots, z_{n+1} = w_0$ in $[z_0, w_0]$ such that

$$|z_j - z_{j+1}| = \frac{|z_0 - w_0|}{n} < \delta$$

for $j = 0, 1, \ldots, n$. Then

$$\begin{aligned} |f(z_0) - f(w_0)| &\leq \sum_{j=0}^{n} |f(z_j) - f(z_{j+1})| \\ &\leq L \sum_{j=0}^{n} |z_j - z_{j+1}| = L|z_0 - w_0| \end{aligned}$$

by (9.4.11). □

REMARK 9.4.12. The conclusion in Lemma 9.4.10 also follows with the hypothesis that f is locally L-lipschitz replaced by the seemingly weaker assumption that

$$\limsup_{z \to w} \frac{|f(z) - f(w)|}{|z - w|} \leq L$$

for each $w \in E$.

See John [**91**] and Nevanlinna [**138**].

LEMMA 9.4.13 (John [**91**]). *If $f : \mathbf{B}(z_0, r) \to \mathbf{R}^2$ is locally L-bilipschitz in $\mathbf{B}(z_0, r)$, then f is L-bilipschitz in $\mathbf{B}(z_0, r/L^2)$.*

PROOF. We may assume without loss of generality that $f(z_0) = z_0 = 0$ and $r = 1$. Then Lemma 9.4.10 implies that f is L-lipschitz in $\mathbf{B}$ and hence that

$$f(\mathbf{B}(0, L^{-2})) \subset \mathbf{B}(0, L^{-1}).$$

To complete the proof, we shall define a continuous function g on $\mathbf{B}(0, L^{-1})$ such that

$$f(g(w)) = w \quad \text{for} \quad w \in \mathbf{B}(0, L^{-1}), \tag{9.4.14}$$

$$g(f(z)) = z \quad \text{for} \quad z \in \mathbf{B}(0, L^{-2}), \tag{9.4.15}$$

for if g is continuous, (9.4.14) will imply that

$$\frac{1}{L} \leq \frac{|f(g(w_1)) - f(g(w_2))|}{|g(w_1) - g(w_2)|} = \frac{|w_1 - w_2|}{|g(w_1) - g(w_2)|}$$

locally in $\mathbf{B}(0, L^{-1})$ and hence that g is L-lipschitz in $\mathbf{B}(0, L^{-1})$ by a second application of Lemma 9.4.10, and with (9.4.15) we will conclude

$$\frac{1}{L} \leq \frac{|f(z_1) - f(z_2)|}{|g(f(z_1)) - g(f(z_2))|} = \frac{|f(z_1) - f(z_2)|}{|z_1 - z_2|}$$

when $z_1, z_2 \in \mathbf{B}(0, L^{-2})$ and hence that f is L-bilipschitz in $\mathbf{B}(0, L^{-2})$.

To establish (9.4.14), we show first that there is a unique continuous function g defined on the radial segments of $\mathbf{B}(0, L^{-1})$ such that $g(0) = 0$ and $f \circ g(z) = z$. For this, given a point w with $|w| = 1$, we define

$$\tau = \tau(w) = \sup t$$

where the supremum is taken over all $t > 0$ for which there exists a continuous function g_t defined on the radial segment $[0, tw]$ with $g_t(0) = 0$ and $f \circ g_t(z) = z$ for $z \in [0, tw]$.

The function g_t is unique for all $t < \tau$. For otherwise there would exist for some t, $0 < t < \tau$, continuous functions $g_{1,t}$ and $g_{2,t}$ defined on $[0, tw]$ so that both are inverses of f with $g_{1,t}(0) = g_{2,t}(0) = 0$. We may assume without loss of generality that $w = 1$ and let

$$s_0 = \inf\{s > 0 : g_{1,t}(s) \neq g_{2,t}(s)\}.$$

Then $0 < s_0 < \tau$ and $g_{1,t}(s_0) = g_{2,t}(s_0)$, since $g_{1,t}$ and $g_{2,t}$ are continuous. Moreover there is a sequence $s_n \to s_0$ so that

$$g_{1,t}(s_n) \neq g_{2,t}(s_n)$$

for all n and

$$\lim_{n\to\infty} g_{1,t}(s_n) = g_{1,t}(s_0) = \lim_{n\to\infty} g_{2,t}(s_n).$$

Since

$$f(g_{1,t}(s_n)) = f(g_{2,t}(s_n)),$$

this contradicts the fact that f is locally injective at $g_{1,t}(s_0)$. Thus the function g_t is a function g independent of the choice of parameter t.

Next the fact that f is a local homeomorphism implies that $\tau > 0$ and that τ is not attained. Hence

$$\lim_{t\to\tau} g(t\,w)$$

either does not exist or exists and is a point in $\partial\mathbf{B}$ by virtue of our normalization. By Lemma 9.4.10 applied to g on $[0, tw]$,

$$|g(t\,w) - g(s\,w)| \leq L(t - s) \qquad \text{for} \qquad 0 \leq s \leq t < \tau.$$

Hence if $t_n \to \tau$, $g(t_n w)$ is a Cauchy sequence and we conclude

$$\lim_{n\to\infty} g(t_n w) = z \notin \mathbf{B} \qquad \text{and} \qquad 1 \leq |z| \leq \tau L$$

which implies that $\tau \geq 1/L$.

Finally, we use the functions $g(t)$ defined on the radii $[0, \tau w)$ above to define a function on $\mathbf{B}(0, L^{-1})$ which we also denote by g. Then since f is a local homeomorphism and a continuous inverse map is determined on a segment once it is given at one point, we conclude that g coincides locally with a continuous map and hence is continuous. Moreover (9.4.15) follows by an argument similar to the one used to prove that the mappings g_t were unique. □

We are now in a position to prove the main result of this section. For this we modify a global approximation argument due to John [**90**] and employ the simple criterion for injectivity introduced by Martio and Sarvas in [**123**].

THEOREM 9.4.16. *If D is a uniform domain with constant c, then D is rigid with $L(D) = L(c) > 1$.*

PROOF. By hypothesis, there exists a constant $c \geq 1$ such that each pair of points $z_1, z_2 \in D$ can be joined by an arc $\gamma \subset D$ such that

$$\begin{aligned} &\text{length}(\gamma) \leq c|z_1 - z_2|, \\ &\min_{j=1,2} \text{length}(\gamma_j) \leq c\,\text{dist}(z, \partial D), \end{aligned} \tag{9.4.17}$$

where γ_1, γ_2 are the components of $\gamma\backslash\{z\}$. We will show that there exists a constant $L(c) > 1$ such that each function f locally L-bilipschitz in D with $1 < L < L(c)$ is injective; we may assume $L > 1$ since otherwise f is trivially injective. We do this by proving that for any such function and any two points $z_1, z_2 \in D$, there is an isometry g of $\mathbf{R}^2$ such that

$$|f(z_j) - g(z_j)| < \frac{|z_1 - z_2|}{2} \tag{9.4.18}$$

for $j = 1, 2$. From this it follows that $f(z_1) \neq f(z_2)$ since otherwise we would have

$$|z_1 - z_2| = |g(z_1) - g(z_2)| \leq |f(z_1) - g(z_1)| + |f(z_2) - g(z_2)| < |z_1 - z_2|.$$

To establish (9.4.18), we join the points z_1, z_2 by the arc γ which satisfies (9.4.17) and let z_0 be the midpoint of γ, i.e.,

$$\text{length}(\gamma(z_1, z_0)) = \text{length}(\gamma(z_2, z_0)) = l$$

where $\gamma(z, z_0)$ denotes the subarc with z_0, z as endpoints. Let $z = z(s)$ be the arclength representation for $\gamma(z_1, z_0)$ with $z(0) = z_0$ and $z(l) = z_1$. Then

$$\text{dist}(z(s), \partial D) \geq \frac{l - s}{c}$$

by (9.4.17). Set

$$q = \frac{cL^2}{1 + cL^2} \in (\tfrac{1}{2}, 1)$$

and for $j = 0, 1, 2, \ldots$ let

$$s_j = (1 - q^j)\, l, \quad \zeta_j = z(s_j), \quad d_j = \frac{q^j}{c}\, l, \quad r_j = \frac{d_j}{L^2}. \tag{9.4.19}$$

Then

$$|\zeta_{j+1} - \zeta_j| \leq s_{j+1} - s_j = (1 - q)q^j\, l = \frac{q^{j+1}\, l}{c\, L^2} = r_{j+1} \tag{9.4.20}$$

and $\text{dist}(\zeta_j, \partial D) > d_j = L^2\, r_j$.

By Lemmas 9.4.1 and 9.4.13 there exists for each j an isometry $g_j : \mathbf{R}^2 \to \mathbf{R}^2$ such that

$$|f(z) - g_j(z)| \leq a(L - 1)\, r_j \tag{9.4.21}$$

in $B_j = \overline{\mathbf{B}}(\zeta_j, r_j)$ where $a < 20$,

$$g_j(z) = E_j(z - \zeta_j) + f(\zeta_j), \tag{9.4.22}$$

and $E_j(z) = e^{i\,\theta_j} z$ or $E_j(z) = e^{i\,\theta_j}\overline{z}$. Next since

$$|\zeta_{j+1} - \zeta_j| \leq r_{j+1},$$

$B_j \cap B_{j+1}$ contains a closed disk B^* of radius $\frac{1}{2}\, r_j$ in which

$$\begin{aligned} |g_j(z) - g_{j+1}(z)| &\leq |f(z) - g_j(z)| + |f(z) - g_{j+1}(z)| \\ &\leq a(L - 1)(r_j + r_{j+1}) \end{aligned} \tag{9.4.23}$$

by (9.4.21). Then $F(z) = E_j(z) - E_{j+1}(z)$ is linear in x and y where $z = x + i\,y$ and $F(B^*)$ is contained in a closed disk of radius

$$t = a(L-1)(r_j + r_{j+1}).$$

Hence

$$|E_j(z) - E_{j+1}(z)| = |F(z)| \le \frac{2\,t}{r_j}|z| = 2\,a(L-1)(1+q)|z| \tag{9.4.24}$$

for $z \in \mathbf{R}^2$. Next $\zeta_j \in B_j \cap B_{j+1}$ and inequality (9.4.23) with $z = \zeta_j$ yields

$$|g_j(\zeta_j) - g_{j+1}(\zeta_j)| \le a(L-1)(1+q)q^j\, r_0. \tag{9.4.25}$$

Finally, if $n > j$, then

$$|\zeta_n - \zeta_j| \le s_n - s_j = (q^j - q^n)\,l \le q^j\,l$$

and we obtain

$$\begin{aligned} |g_j(\zeta_n) - g_{j+1}(\zeta_n)| &\le |g_j(\zeta_j) - g_{j+1}(\zeta_j)| + |E_j(\zeta_n - \zeta_j) - E_{j+1}(\zeta_n - \zeta_j)| \\ &\le a(L-1)(1+q)q^j(r_0 + 2\,l) \end{aligned}$$

from (9.4.22), (9.4.24), and (9.4.25). This, together with (9.4.19), implies that

$$\begin{aligned} |f(\zeta_n) - g_0(\zeta_n)| &= |g_n(\zeta_n) - g_0(\zeta_n)| \\ &\le \sum_{j=0}^{n-1} |g_j(\zeta_n) - g_{j+1}(\zeta_n)| \\ &\le a(L-1)\frac{1+q}{1-q}(r_0 + 2\,l) \\ &= a(L-1)\frac{(r_0 + 2\,l)^2}{r_0} \\ &\le 9\,ac\,L^2(L-1)\,l. \end{aligned}$$

Here $\zeta_n \to z_1$ as $n \to \infty$ and we conclude that

$$|f(z_1) - g_0(z_1)| \le 9a\,c\,L^2(L-1)\,\mathrm{length}(\gamma(z_1, z_0))$$

and hence with (9.4.17) that

$$|f(z_1) - g_0(z_1)| \le 9a\,c^2\,L^2(L-1)\frac{|z_1 - z_2|}{2} < \frac{|z_1 - z_2|}{2}$$

when $1 < L < L(c)$. Finally,

$$|f(z_2) - g_0(z_2)| < \frac{|z_1 - z_2|}{2}$$

by the above argument applied to z_0, z_2 and $f(z_1) \ne f(z_2)$. □

9.5. Rigid domains are linearly locally connected

We show next that rigid domains are linearly locally connected. Our proof makes use of the following technical lemma concerning a special class of bilipschitz maps.

LEMMA 9.5.1. *Suppose that $\phi(t)$ is a real-valued function defined for $0 < t < \infty$, that*

$$|\phi(t_1) - \phi(t_2)| \le b\,|\log t_1 - \log t_2| \tag{9.5.2}$$

for $0 < t_1, t_2 < \infty$, and that

$$f(z) = \begin{cases} z\, e^{i\,\phi(|z|)} & \text{if } \ 0 < |z| < \infty, \\ 0 & \text{if } \ z = 0. \end{cases} \tag{9.5.3}$$

Then f is $(1+b)$-bilipschitz in $\mathbf{R}^2$.

PROOF. Choose distinct points $z_1, z_2 \in \mathbf{R}^2$ with $|z_1| \le |z_2|$. If $z_1 \ne 0$, then

$$\begin{aligned} |f(z_1) - f(z_2)| &\le |z_1 - z_2| + |z_1||e^{i\,\phi(|z_1|)} - e^{i\,\phi(|z_2|)}| \\ &\le |z_1 - z_2| + |z_1||\phi(|z_1|) - \phi(|z_2|)| \\ &\le |z_1 - z_2| + b\,|z_1||\log|z_1| - \log|z_2|| \\ &\le (1+b)|z_1 - z_2| \end{aligned}$$

by (9.5.2) while

$$|f(z_1) - f(z_2)| = |z_2| \le (1+b)|z_1 - z_2|$$

if $z_1 = 0$. Next since f^{-1} is given by (9.5.3) with $-\phi$ in place of ϕ, the above argument can be applied to f^{-1} to complete the proof. □

THEOREM 9.5.4 (Gehring [**50**]). *Suppose that D is a domain in $\mathbf{R}^2$ with $L(D) > 1$ and that c is a constant where*

$$\log c > \frac{\pi}{L(D) - 1}. \tag{9.5.5}$$

Then for each $z_0 \in \mathbf{R}^2$ and $0 < r < \infty$, $D \cap \partial\mathbf{B}(z_0, r)$ is empty or lies in a component of

$$G = D \cap (\mathbf{B}(z_0, cr) \setminus \overline{\mathbf{B}}(z_0, r/c)).$$

PROOF. Suppose there exist points $z_1, z_2 \in D \cap \partial\mathbf{B}(z, r)$ which lie in different components G_1, G_2 of G. By making a change of variable we may assume that $z_0 = 0$. Choose $\theta \in [-\pi, \pi]$ so that $z_2 = z_1\, e^{i\,\theta}$ and let f be as in (9.5.3) where

$$\phi(t) = \begin{cases} 0 & \text{if } \ t \notin [r/c, cr], \\[2ex] \log \dfrac{c\,t}{r}\, \dfrac{\theta}{\log c} & \text{if } \ t \in [r/c, r], \\[2ex] \log \dfrac{c\,r}{t}\, \dfrac{\theta}{\log c} & \text{if } \ t \in [r, cr]. \end{cases}$$

Then ϕ satisfies (9.5.2) with $b = \pi/\log c$ and f is L-bilipschitz in $\mathbf{R}^2$ by Lemma 9.5.1, where

$$L = 1 + \frac{\pi}{\log c} < L(D). \tag{9.5.6}$$

Set

$$g(z) = \begin{cases} z & \text{if } \ z \in D \setminus G_1, \\ f(z) & \text{if } \ z \in G_1. \end{cases}$$

If U is any open disk in D, then either $U \subset D \setminus G_1$, in which case $g(z) = z$ in U, or

$$U \cap G \subset G_1, \qquad \text{whence} \qquad U \subset G_1 \cup (D \setminus G),$$

in which case $g(z) = f(z)$ in U. Hence g is locally L-bilipschitz in D and hence injective by (9.5.6). On the other hand, $z_2 \notin G_1$,

$$g(z_2) = z_2 = z_1\, e^{i\,\theta} = z_1\, e^{i\,\phi(|z_1|)} = g(z_1),$$

and we have a contradiction. □

The desired result is now an easy consequence of Theorem 9.5.4.

COROLLARY 9.5.7. *If D is a rigid domain in $\mathbf{R}^2$, then D is linearly locally connected with constant c where*

$$\log c > \frac{\pi}{L(D) - 1}. \tag{9.5.8}$$

PROOF. Suppose that z_1 and z_2 are points in $D \cap \mathbf{B}(z_0, r)$ and that α is an arc joining z_1, z_2 in D. If $\alpha \not\subset \mathbf{B}(z_0, cr)$, let w_1 and w_2 be the first and last points where α meets $\partial\mathbf{B}(z_0, r)$. Theorem 9.5.4 then implies that w_1 and w_2 can be joined in $D \cap \mathbf{B}(z_0, cr)$. Hence z_1 and z_2 can be joined in $D \cap \mathbf{B}(z_0, cr)$ and

$$D \cap \mathbf{B}(z_0, r) \text{ lies in a component of } D \cap \mathbf{B}(z_0, cr).$$

In the same way we see that

$$D \setminus \overline{\mathbf{B}}(z_0, r) \text{ lies in a component of } D \setminus \overline{\mathbf{B}}(z_0, r/c)$$

and hence that D is linearly locally connected with constant c. □

9.6. Uniform domains have the min-max property

A domain $D \subset \mathbf{R}^2$ is uniform if for some constant $a \geq 1$ each pair of points $z_1, z_2 \in D$ can be joined by a curve $\gamma \subset D$ such that

$$\text{length}(\gamma) \leq a\,|z_1 - z_2|, \tag{9.6.1}$$

$$\min_{j=1,2} \text{length}(\gamma_j) \leq a\, \text{dist}(z, \partial D) \tag{9.6.2}$$

for each $z \in \gamma$, where γ_1, γ_2 are the components of $\gamma \setminus \{z\}$.

The domain D has the min-max property if for some constant $b \geq 1$ each $z_1, z_2 \in D$ can be joined by a curve $\gamma \subset D$ such that

$$\frac{1}{b} \min_{j=1,2} |z_j - w| \leq |z - w| \leq b \max_{j=1,2} |z_j - w| \tag{9.6.3}$$

for each $z \in \gamma$ and each $w \notin D$.

The main result of this section, namely that a uniform domain has the min-max property, is an immediate consequence of the following observation.

REMARK 9.6.4. Suppose that γ is a curve joining z_1, z_2 in a domain $D \subset \mathbf{R}^2$. If γ satisfies (9.6.1) and (9.6.2) for each $z \in \gamma$, then it satisfies (9.6.3) for each $z \in \gamma$ and each $w \notin D$ with $b = a + 1$.

PROOF. Fix $z \in \gamma$ and $w \notin D$. Then

$$\begin{aligned} |z-w| &\le \frac{|z-z_1|+|z_1-w|+|z-z_2|+|z_2-w|}{2} \\ &\le \frac{\text{length}(\gamma)+|z_1-w|+|z_2-w|}{2} \\ &\le \frac{(a+1)(|z_1-w|+|z_2-w|)}{2} \\ &\le (a+1)\max_{j=1,2}|z_j-w| \end{aligned}$$

while

$$\begin{aligned} \min_{j=1,2}|z_j-w| &\le \min_{j=1,2}\text{length}(\gamma_j)+|z-w| \le a\,\text{dist}(z,\partial D)+|z-w| \\ &\le (a+1)|z-w| \end{aligned}$$

where γ_1, γ_2 are the components of $\gamma \setminus \{z\}$. □

9.7. Min-max property and local connectivity

We conclude this sequence of implications by showing that a domain D with the min-max property is linearly locally connected.

THEOREM 9.7.1. *Suppose that for some constant $a \ge 1$ each pair of points z_1, z_2 in a domain $D \subset \mathbf{R}^2$ can be joined by a curve α in D such that*

$$\frac{1}{a}\min_{j=1,2}|z_j-w| \le |z-w| \le a\max_{j=1,2}|z_j-w| \tag{9.7.2}$$

for each $z \in \alpha$ and each $w \notin D$. Then for each $z_0 \in \mathbf{R}^2$ and each $r > 0$

$$D \cap \mathbf{B}(z_0,r) \text{ lies in a component of } D \cap \mathbf{B}(z_0,cr), \tag{9.7.3}$$

$$D \setminus \overline{\mathbf{B}}(z_0,r) \text{ lies in a component of } D \setminus \overline{\mathbf{B}}(z_0,r/c) \tag{9.7.4}$$

where $c = 8a+1$.

PROOF. We shall show that there exists a constant $b = 4a$ such that the points z_1, z_2 can be joined by a curve β such that

$$\text{diam}(\beta) \le b\,|z_1-z_2|, \tag{9.7.5}$$

$$\min_{j=1,2}|z-z_j| \le b\,\text{dist}(z,\partial D) \tag{9.7.6}$$

for each $z \in \beta$, for then (9.7.3) and (9.7.4) will follow from Theorem 8.3.1 with $c = 2b+1$.

If $|z_1-z_2| < \text{dist}(z_1,\partial D)$, let $\beta = [z_1,z_2]$. Then

$$[z_1,z_2] \subset \mathbf{B}(z_1,\text{dist}(z_1,\partial D)) \subset D$$

and (9.7.5) and (9.7.6) hold with $b = 1$.

If $|z_1-z_2| \ge \text{dist}(z_1,\partial D)$, choose $w_1 \in \partial D$ so that

$$|z_1-w_1| = \text{dist}(z_1,\partial D) \le |z_1-z_2|$$

and let $\beta = \alpha$. If $z \in \beta$, then

$$|z-w_1| \le \max_{j=1,2}|z_j-w_1| \le a(|z_1-w_1|+|z_1-z_2|) \le 2a|z_1-z_2|$$

and hence (9.7.5) holds with $b = 4a$. Finally, choose $w \in \partial D$ so that

$$|z-w| = \text{dist}(z,\partial D).$$

Then

$$\min_{j=1,2} |z - z_j| \leq \min_{j=1,2} |z_j - w| + |z - w| \leq (a+1)\text{dist}(z, \partial D)$$

and (9.7.6) holds with $b = a + 1$ and hence with $b = 4a$. □

CHAPTER 10

Third series of implications

We establish here three more characterizations for quasidisks making use of the following result established in Chapter 8 :

D is a quasidisk if and only if it has the hyperbolic segment property.

We prove, in particular, the following two chains of statements for a simply connected domain D.

1° A quasidisk is a BMO-extension domain.
2° The inequality $h_D \leq c\, j_D$ holds in a BMO-extension domain.
3° The inequality $h_D \leq c\, j_D$ implies the segment property.
4° The inequaltiy $h_D \leq c\, a_D$ holds in a quasidisk.
5° The inequality $h_D \leq c\, a_D$ implies the hyperbolic segment property.

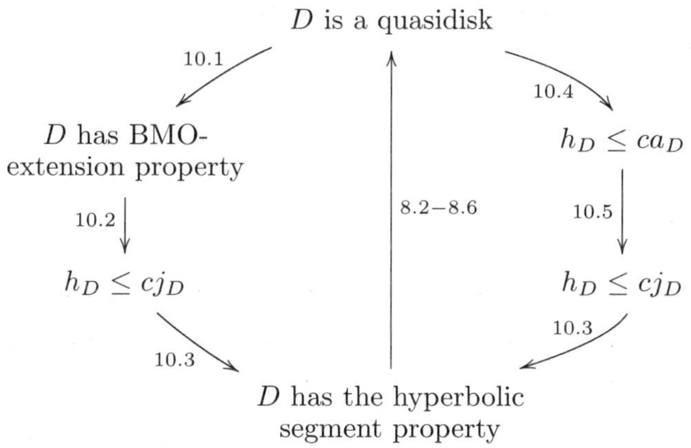

FIGURE 10.1

10.1. Quasidisks and BMO-extension

We shall need the following result due to H. M. Reimann concerning the invariance of functions with bounded mean oscillation under quasiconformal mappings.

LEMMA 10.1.1. *Suppose that D and D' are domains in $\mathbf{R}^2$ and that $f : D \to D'$ is K-quasiconformal. If v is in* $\mathrm{BMO}(D')$, *then $u = v \circ f$ is in* $\mathrm{BMO}(D)$ *and*

$$\|u\|_{\mathrm{BMO}(D)} \leq a \, \|v\|_{\mathrm{BMO}(D')} \, ,$$

where $a = a(K)$.

The proof is based on integrability properties of quasiconformal mappings. See, for example, Theorem 2 in Reimann [**146**] or Theorem V.C.3 in Reimann-Rychener [**147**].

THEOREM 10.1.2 (Jones [**94**]). *If D is a K-quasidisk, then D is a* BMO-*extension domain with constant $c = c(K)$.*

PROOF. By hypothesis there is a K-quasiconformal self-mapping f of $\overline{\mathbf{R}}^2$ which maps D onto a disk or half-plane D'. By composing f with a Möbius transformation we may assume that $f(\infty) = \infty$.

Suppose now that u is in $\mathrm{BMO}(D)$ and let $u' = u \circ f^{-1}$. Then by Lemma 10.1.1, u' is in $\mathrm{BMO}(D')$ and

$$\|u'\|_{\mathrm{BMO}(D')} \leq a \, \|u\|_{\mathrm{BMO}(D)}$$

where $a = a(K)$. Next by Theorem 5.1.7, u' has an extension v' in $\mathrm{BMO}(\mathbf{R}^2)$ with

$$\|v'\|_{\mathrm{BMO}(\mathbf{R}^2)} \leq b \, \|u'\|_{\mathrm{BMO}(D')} \, ,$$

where b is an absolute constant. Then again by Lemma 10.1.1, $v = v' \circ f$ is in $\mathrm{BMO}(\mathbf{R}^2)$ with

$$\|v\|_{\mathrm{BMO}(\mathbf{R}^2)} \leq a \, \|v'\|_{\mathrm{BMO}(\mathbf{R}^2)} \, .$$

Thus v is the desired BMO-extension of u and

$$\|v\|_{\mathrm{BMO}(\mathbf{R}^2)} \leq c \, \|u\|_{\mathrm{BMO}(D)} \, ,$$

where $c = a(K)^2 b$. □

10.2. BMO-extension and the hyperbolic metric

We show next that if D is a BMO-extension domain, then the hyperbolic metric $h_D(z_1, z_2)$ in D is bounded above by a constant times the distance-ratio metric

$$j_D(z_1, z_2) = \log\left(\frac{|z_1 - z_2|}{\mathrm{dist}(z_1, \partial D)} + 1\right)\left(\frac{|z_1 - z_2|}{\mathrm{dist}(z_2, \partial D)} + 1\right).$$

We begin with the following preliminary result.

LEMMA 10.2.1 (Gehring-Hag [**56**]). *Suppose that D is simply connected and that there exist constants $a > 0$ and $b > 0$ such that*

$$h_D(z_1, z_2) \leq a \, j_D(z_1, z_2) + b$$

for $z_1, z_2 \in D$. Then

$$h_D(z_1, z_2) \leq c \, j_D(z_1, z_2) \tag{10.2.2}$$

for $z_1, z_2 \in D$ where $c = \max(a + b, 2)$.

PROOF. Choose $z_1, z_2 \in D$ with

$$r_1 = \text{dist}(z_1, \partial D) \le \text{dist}(z_2, \partial D) = r_2$$

and let $t = j_D(z_1, z_2)$. If $1 \le t < \infty$, then

$$h_D(z_1, z_2) \le \left(a + \frac{b}{t}\right) j_D(z_1, z_2) \le (a+b)\, j_D(z_1, z_2).$$

If $0 < t < 1$, then

$$\left(\frac{|z_1 - z_2|}{r_2} + 1\right)^2 \le e^t \quad \text{and} \quad s = \frac{|z_1 - z_2|}{r_2} \le e^{t/2} - 1 < 1.$$

Thus $z_1, z_2 \in D' = \mathbf{B}(z_2, r_2) \subset D$ and

$$h_D(z_1, z_2) \le h_{D'}(z_1, z_2) = \log\left(\frac{1+s}{1-s}\right) \le t + t^2 < 2\, j_D(z_1, z_2).$$

Inequality (10.2.2) then follows from what was proved above. □

THEOREM 10.2.3 (Jones [**94**]). *If D has the* BMO*-extension property with constant a, then*

$$h_D(z_1, z_2) \le c\, j_D(z_1, z_2) \tag{10.2.4}$$

for $z_1, z_2 \in D$ where $c = c(a)$.

PROOF. Fix $z_1, z_2 \in D$ and let

$$u(z) = h_D(z, z_1)$$

for $z \in D$. Then by Lemma 5.1.2, u is in BMO(D) with

$$\|u\|_{\text{BMO}(D)} \le 4.$$

Next, by hypothesis, u has an extension v in BMO$(\mathbf{R}^2)$ with

$$\|v\|_{\text{BMO}(\mathbf{R}^2)} \le a\, \|u\|_{\text{BMO}(D)} \le 4\, a. \tag{10.2.5}$$

For $j = 1, 2$ let $B_j = \mathbf{B}(z_j, r_j)$ where $r_j = \text{dist}(z_j, \partial D)$. By relabeling if necessary we may assume that $r_1 \le r_2$. Next let

$$B_0 = \mathbf{B}(z_2, r_0) \quad \text{where} \quad r_0 = |z_1 - z_2| + r_1.$$

Then

$$|z - z_2| \le |z - z_1| + |z_1 - z_2| < r_0$$

if $z \in B_1$,

$$|z - z_2| < \text{dist}(z_2, \partial D) \le |z_1 - z_2| + \text{dist}(z_1, \partial D) = r_0$$

if $z \in B_2$, and $B_j \subset B_0$ for $j = 1, 2$. Hence

$$|v_{B_j} - v_{B_0}| \le 2\, a\, e \left(\log \frac{m(B_0)}{m(B_j)} + 1\right) \le 4\, a\, e \log\left(\frac{|z_1 - z_2|}{r_j} + 1\right) + 2\, a\, e$$

by Lemma 5.1.5 and we obtain

$$|u_{B_1} - u_{B_2}| = |v_{B_1} - v_{B_2}| \le b\, j_D(z_1, z_2) + b$$

where $b = 4\, a\, e$. Finally,

$$|u_{B_j} - u(z_j)| \le 2$$

for $j = 1, 2$ by (5.1.3) in the proof of Lemma 5.1.2. Thus

$$\begin{aligned} h_D(z_1, z_2) &= |u(z_1) - u(z_2)| \\ &\leq |u(z_1) - u_{B_1}| + |u_{B_1} - u_{B_2}| + |u_{B_2} - u(z_2)| \\ &\leq b\, j_D(z_1, z_2) + b + 4 \end{aligned}$$

and we obtain (10.2.4) with $c = 2b + 4$ from Lemma 10.2.1. □

10.3. Hyperbolic metric and hyperbolic segments

We begin with the following lower bound for the hyperbolic distance $h_D(z_1, z_2)$ similar to that given in Lemma 3.3.5.

LEMMA 10.3.1 (Gehring-Palka [**68**]). *If D is simply connected, then*

$$\left| \log \frac{\operatorname{dist}(z_1, \partial D)}{\operatorname{dist}(z_2, \partial D)} \right| \leq 2\, h_D(z_1, z_2) \tag{10.3.2}$$

for $z_1, z_2 \in D$.

PROOF. As in the proof of Lemma 3.3.5 we see that

$$\begin{aligned} \log \left(\frac{\operatorname{dist}(z_2, \partial D)}{\operatorname{dist}(z_1, \partial D)} \right) &\leq \log \left(\frac{|z_1 - z_2| + \operatorname{dist}(z_1, \partial D)}{\operatorname{dist}(z_1, \partial D)} \right) \\ &= \log \left(\frac{|z_1 - z_2|}{\operatorname{dist}(z_1, \partial D)} + 1 \right) \leq 2\, h_D(z_1, z_2). \end{aligned}$$

Interchanging z_1 and z_2 yields

$$\log \left(\frac{\operatorname{dist}(z_1, \partial D)}{\operatorname{dist}(z_2, \partial D)} \right) \leq 2\, h_D(z_1, z_2)$$

and we obtain (10.3.2). □

We complete the first chain of implications in this chapter by establishing the following result.

THEOREM 10.3.3 (Gehring-Osgood [**67**]). *Suppose that D is simply connected and that*

$$h_D(z_1, z_2) \leq a\, j_D(z_1, z_2) \tag{10.3.4}$$

for each $z_1, z_2 \in D$ where a is a constant. Then for each hyperbolic segment γ joining $z_1, z_2 \in D$ and each $z \in \gamma$,

$$\begin{aligned} &\operatorname{length}(\gamma) \leq c\, |z_1 - z_2|, \\ &\min_{j=1,2} \operatorname{length}(\gamma_j) \leq c \operatorname{dist}(z, \partial D), \end{aligned} \tag{10.3.5}$$

where γ_1, γ_2 are the components of $\gamma \setminus \{z\}$ and $c = c(a)$.

PROOF. Fix $z_1, z_2 \in D$, let γ be the hyperbolic segment joining z_1 to z_2 in D, and set

$$r = \min \left(\sup_{z \in \gamma} \operatorname{dist}(z, \partial D),\ 2\, |z_1 - z_2| \right). \tag{10.3.6}$$

We shall consider the cases where

$$r < \max_{j=1,2}(\operatorname{dist}(z_j, \partial D)), \tag{10.3.7}$$

$$r \geq \max_{j=1,2}(\operatorname{dist}(z_j, \partial D)) \tag{10.3.8}$$

separately.

Case where $r < \sup_{z\in\gamma} \operatorname{dist}(z, \partial D)$

If $r < \operatorname{dist}(z_1, \partial D)$, then $r = 2\,|z_1 - z_2|$,

$$2\,|z_1 - z_2| < \operatorname{dist}(z_1, \partial D) \leq 2 \operatorname{dist}(z, \partial D)$$

for z on the segment $\beta = [z_1, z_2] \subset D$, and

$$h_D(z_1, z_2) \leq \int_\beta \rho_D(z)\, ds \leq \int_\beta \frac{2}{\operatorname{dist}(z, \partial D)}\, ds \leq \frac{4|z_1 - z_2|}{\operatorname{dist}(z_1, \partial D)} \leq 2.$$

Since

$$h_D(z_1, z) \leq h_D(z_1, z_2)$$

for $z \in \gamma$, Lemma 10.3.1 implies that

$$e^{-4} \operatorname{dist}(z_1, \partial D) \leq \operatorname{dist}(z, \partial D) \leq e^4 \operatorname{dist}(z_1, \partial D)$$

for $z \in \gamma$. Thus

$$\begin{aligned} \operatorname{length}(\gamma) &\leq e^4 \operatorname{dist}(z_1, \partial D) \int_\gamma \frac{ds}{\operatorname{dist}(z, \partial D)} \\ &\leq e^4 \operatorname{dist}(z_1, \partial D) \int_\gamma 2\,\rho_D(z)\, ds \\ &= 2e^4 \operatorname{dist}(z_1, \partial D)\, h_D(z_1, z_2) \leq 8e^4\, |z_1 - z_2| \end{aligned} \tag{10.3.9}$$

and

$$\begin{aligned} \operatorname{length}(\gamma(z_1, z)) &\leq 8e^4\, |z_1 - z_2| \leq 4e^4 \operatorname{dist}(z_1, \partial D) \\ &\leq 4e^8 \operatorname{dist}(z, \partial D) \end{aligned} \tag{10.3.10}$$

for $z \in \gamma$. If $r < \operatorname{dist}(z_2, \partial D)$, then we obtain (10.3.9) and (10.3.10) by reversing the roles of z_1 and z_2 in the above argument. Hence (10.3.7) implies (10.3.5) with $c = 4e^8$.

Case where $r \geq \sup_{z\in\gamma} \operatorname{dist}(z, \partial D)$

In this case there exists $z_0 \in \gamma$ such that

$$\operatorname{dist}(z_0, \partial D) = \sup_{z\in\gamma} \operatorname{dist}(z, \partial D) \geq r.$$

For $j = 1, 2$ let m_j be the largest integer for which

$$2^{m_j} \operatorname{dist}(z_j, \partial D) \leq r$$

and let w_j be the first point of $\gamma(z_j, z_0)$ with

$$\operatorname{dist}(w_j, \partial D) = 2^{m_j} \operatorname{dist}(z_j, \partial D)$$

as we traverse γ from z_j towards z_0. Then

$$\operatorname{dist}(w_j, \partial D) \leq r < 2 \operatorname{dist}(w_j, \partial D). \tag{10.3.11}$$

We shall show first that

$$\begin{aligned}\text{length}(\gamma(z_j, w_j)) &\leq b\,\text{dist}(w_j, \partial D),\\ \text{length}(\gamma(z_j, z)) &\leq be^{b/2}\text{dist}(z, \partial D)\end{aligned} \tag{10.3.12}$$

for $j = 1, 2$ and $z \in \gamma(z_j, w_j)$ where $b = 64\,a^2$. We need only consider the case where $j = 1$ and $m_1 \geq 1$.

Choose points $\zeta_1, \zeta_2, \ldots, \zeta_{m_1+1} \in \gamma(z_1, w_1)$ so that $\zeta_1 = z_1$ and ζ_k is the first point of $\gamma(z_1, w_1)$ with

$$\text{dist}(\zeta_k, \partial D) = 2^{k-1}\,\text{dist}(z_1, \partial D) \tag{10.3.13}$$

as we traverse γ from z_1 towards w_1. Then $\zeta_{m_1+1} = w_1$. Fix k, $1 \leq k \leq m_1$, and let

$$t = \frac{\text{length}(\gamma(\zeta_k, \zeta_{k+1}))}{\text{dist}(\zeta_k, \partial D)} \geq \frac{|\zeta_k - \zeta_{k+1}|}{\text{dist}(\zeta_k, \partial D)}.$$

If $z \in \gamma(\zeta_k, \zeta_{k+1})$, then

$$\text{dist}(z, \partial D) \leq \text{dist}(\zeta_{k+1}, \partial D) = 2\,\text{dist}(\zeta_k, \partial D)$$

and

$$t \leq 2\int_{\gamma_k} \frac{ds}{\text{dist}(z, \partial D)} \leq 4\,h_D(\zeta_k, \zeta_{k+1}),$$

where $\gamma_k = \gamma(\zeta_k, \zeta_{k+1})$. Next

$$j_D(\zeta_k, \zeta_{k+1}) \leq 2\log\left(\frac{|\zeta_k - \zeta_{k+1}|}{\text{dist}(\zeta_k, \partial D)} + 1\right) \leq 2\,\log(t+1) \leq 2\sqrt{t}$$

since the function

$$f(x) = \sqrt{x} - \log(x+1)$$

is increasing for $x > 0$ with $f(0) = 0$. Hence

$$t \leq 4\,h_D(\zeta_k, \zeta_{k+1}) \leq 4a\,j_D(\zeta_k, \zeta_{k+1}) \leq 8a\sqrt{t}$$

by (10.3.4), whence $t \leq 64\,a^2 = b$ and

$$h_D(\zeta_k, \zeta_{k+1}) \leq 2a\sqrt{t} \leq b/4.$$

Next if $z \in \gamma(\zeta_k, \zeta_{k+1})$, then

$$0 < \log\left(\frac{\text{dist}(\zeta_{k+1}, \partial D)}{\text{dist}(z, \partial D)}\right) \leq 2\,h_D(z, \zeta_{k+1}) \leq 2\,h_D(\zeta_k, \zeta_{k+1}) \leq b/2$$

by Lemma 10.3.1. We conclude that

$$\begin{aligned}\text{length}(\gamma(\zeta_k, \zeta_{k+1})) &\leq b\,\text{dist}(\zeta_k, \partial D),\\ \text{dist}(\zeta_{k+1}, \partial D) &\leq e^{b/2}\text{dist}(z, \partial D)\end{aligned} \tag{10.3.14}$$

for $k = 1, 2, \ldots, m_1$ and $z \in \gamma(\zeta_k, \zeta_{k+1})$.

We now complete the proof for the inequalities in (10.3.12) as follows. By (10.3.13) and (10.3.14)

$$\begin{aligned}\text{length}(\gamma(z_1, w_1)) &= \sum_{k=1}^{m_1} \text{length}(\gamma(\zeta_k, \zeta_{k+1})) \leq b\sum_{k=1}^{m_1} \text{dist}(\zeta_k, \partial D)\\ &= b\,(2^{m_1} - 1)\,\text{dist}(z_1, \partial D) < b\,\text{dist}(w_1, \partial D).\end{aligned}$$

This is the first inequality. For the second, if $z \in \gamma(z_1, w_1)$, then $z \in \gamma(\zeta_k, \zeta_{k+1})$ for some k, $1 \leq k \leq m_1$, and

$$\begin{aligned}\operatorname{length}(\gamma(z_1, z)) &\leq \sum_{j=1}^{k} \operatorname{length}(\gamma(\zeta_k, \zeta_{k+1})) \leq b \sum_{j=1}^{k} \operatorname{dist}(\zeta_k, \partial D) \\ &= b(2^k - 1) \operatorname{dist}(z_1, \partial D) < b \operatorname{dist}(\zeta_{k+1}, \partial D) \\ &\leq b\, e^{b/2} \operatorname{dist}(z, \partial D)\end{aligned}$$

again by (10.3.13) and (10.3.14).

We show next that if $\operatorname{dist}(w_1, \partial D) \leq \operatorname{dist}(w_2, \partial D)$, then

$$\begin{aligned}\operatorname{length}(\gamma(w_1, w_2)) &\leq 2be^{2b} \operatorname{dist}(w_1, \partial D), \\ \operatorname{dist}(w_2, \partial D) &\leq e^{2b} \operatorname{dist}(z, \partial D)\end{aligned} \tag{10.3.15}$$

for all $z \in \gamma(w_1, w_2)$. We may assume that $w_1 \neq w_2$ since otherwise there is nothing to prove. By hypothesis

$$r \geq \max_{j=1,2}(\operatorname{dist}(z_j, \partial D)),$$

where r is as in (10.3.6), and we have the following two possible subcases:

$$r = \sup_{z \in \gamma} \operatorname{dist}(z, \partial D), \tag{10.3.16}$$

$$r = 2\,|z_1 - z_2|. \tag{10.3.17}$$

If (10.3.16) holds, set

$$t = \frac{\operatorname{length}(\gamma(w_1, w_2))}{\operatorname{dist}(w_1, \partial D)}.$$

Then

$$\operatorname{dist}(z, \partial D) \leq r < 2 \operatorname{dist}(w_1, \partial D)$$

if $z \in \gamma(w_1, w_2)$ by (10.3.11) and we can repeat the proof for the first part of (10.3.14), with w_1 in place of ζ_k and w_2 in place of ζ_{k+1}, to obtain (10.3.15).

Next if (10.3.17) holds, then

$$\begin{aligned}|w_1 - w_2| &\leq \sum_{j=1,2} \operatorname{length}(\gamma(z_j, w_j)) + |z_1 - z_2| \\ &\leq b \sum_{j=1,2} \operatorname{dist}(w_j, \partial D) + r/2 \\ &\leq 4b \operatorname{dist}(w_1, \partial D)\end{aligned}$$

by (10.3.11) and the first part of (10.3.12). Hence

$$h_D(w_1, w_2) \leq a\, j_D(w_1, w_2) \leq 2a \log\left(\frac{|w_1 - w_2|}{\operatorname{dist}(w_1, \partial D)}\right) \leq 2a \log 5b < b.$$

If $z \in \gamma(w_1, w_2)$, then

$$e^{-2b} \operatorname{dist}(w_2, \partial D) \leq \operatorname{dist}(z, \partial D) \leq e^{2b} \operatorname{dist}(w_1 \partial D)$$

by Lemma 10.3.1 and, as in (10.3.9),

$$\begin{aligned}\text{length}(\gamma(w_1, w_2)) &\le e^{2b}\,\text{dist}(w_1, \partial D) \int_{\gamma_1} \frac{ds}{\text{dist}(z, \partial D)} \\ &\le e^{2b}\,\text{dist}(w_1, \partial D) \int_{\gamma_1} 2\,\rho_D(z)\,ds \\ &\le 2b\,e^{2b}\text{dist}(w_1, \partial D)\end{aligned} \tag{10.3.18}$$

where $\gamma_1 = \gamma(w_1, w_2)$, and we again obtain (10.3.15).

We are now in a position to complete the proof of Theorem 10.3.3. By relabeling and (10.3.11) we may assume that

$$\text{dist}(w_1, \partial D) \le \text{dist}(w_2, \partial D) \le 2\,\text{dist}(w_1, \partial D).$$

Then

$$\begin{aligned}\text{length}(\gamma) &\le \sum_{j=1,2} \text{length}(\gamma(z_j, w_j)) + \text{length}(\gamma(w_1, w_2)) \\ &\le \sum_{j=1,2} \text{dist}(w_j, \partial D) + 2b\,e^{2b}\text{dist}(w_1, \partial D) \\ &\le 4b\,e^{2b}\text{dist}(w_2, \partial D) \le 4b\,e^{2b}r \le 4b\,e^{2b}\,|z_1 - z_2|\end{aligned} \tag{10.3.19}$$

by (10.3.12) and (10.3.18). Next if $z \in \gamma$, then either $z \in \gamma(z_j, w_j)$ and

$$\min_{j=1,2} \text{length}(\gamma(z_j, z)) \le \text{length}(\gamma(z_j, z)) \le b\,e^{b/2}\text{dist}(z, \partial D) \tag{10.3.20}$$

or $z \in \gamma(w_1, w_2)$ and

$$\begin{aligned}\min_{j=1,2} \text{length}(\gamma(z_j, z)) &\le \text{length}(\gamma)/2 \le 2b\,e^{2b}\text{dist}(w_2, \partial D) \\ &\le 2b\,e^{4b}\text{dist}(z, \partial D)\end{aligned} \tag{10.3.21}$$

by (10.3.15). Hence (10.3.5) follows from (10.3.19), (10.3.20), and (10.3.21) with $c = 4b\,e^{4b}$. □

10.4. Apollonian metric in a quasidisk

If D is a simply connected domain, then

$$a_D(z_1, z_2) \le 4\,h_D(z_1, z_2)$$

for $z_1, z_2 \in D$ by Lemma 3.3.5. We show here that this inequality can be reversed if D is a K-quasidisk, namely that

$$h_D(z_1, z_2) \le c\,a_D(z_1, z_2)$$

for $z_1, z_2 \in D$ where $c = c(K)$. We begin with the following result.

LEMMA 10.4.1. *Suppose that $K \ge 1$ and $c \ge \frac{1}{2}$. If $u > 0$ and*

$$v + 1 \le c^{K-1}(u+1)^K, \tag{10.4.2}$$

then

$$v \le (2c)^{2(K-1)}\,\max(u^K, u^{-K}). \tag{10.4.3}$$

PROOF. Set

$$f(u) = c^{K-1}(u+1)^K - 1 - (2^K c^{K-1} - 1)u^K.$$

Then

$$f'(u) = K(cu)^{K-1} g(u)$$

where

$$g(u) = (1+u^{-1})^{K-1} - 2^K + c^{1-K}.$$

Hence

$$g(u) \le g(1) \le 0, \qquad f'(u) \le 0, \qquad f(u) \le f(1) = 0$$

and, with (10.4.2),

$$v \le f(u) + (2^K c^{K-1} - 1)u^K \le (2^K c^{K-1} - 1)u^K \le (2c)^{2(K-1)} u^K$$

for $u \ge 1$. This implies (10.4.3) when $u \ge 1$. The same argument with u^{-1} in place of u yields (10.4.3) when $0 < u \le 1$. □

THEOREM 10.4.4. *If f is a K-quasiconformal self-mapping of $\overline{\mathbf{R}}^2$ and if D is a proper subdomain of $\overline{\mathbf{R}}^2$, then*

$$a_{f(D)}(f(z_1), f(z_2)) \le K\, a_D(z_1, z_2) + b(K) \tag{10.4.5}$$

for $z_1, z_2 \in D$ where

$$b(K) = 2(K-1)\log 32. \tag{10.4.6}$$

PROOF. By performing preliminary Möbius transformations we may assume that $D, D' \subset \mathbf{R}^2$. Fix $z_1, z_2 \in D$ and choose $w_1, w_2 \in \partial D$ so that

$$a_{f(D)}(f(z_1), f(z_2)) = \log \frac{|f(z_1) - f(w_1)||f(z_2) - f(w_2)|}{|f(z_1) - f(w_2)||f(z_2) - f(w_1)|}.$$

Then by Theorem 1.3.10,

$$v + 1 \le 16^{K-1}(u+1)^K$$

where

$$v = \frac{|f(z_1) - f(w_1)||f(z_2) - f(w_2)|}{|f(z_1) - f(w_2)||f(z_2) - f(w_1)|}, \qquad u = \frac{|z_1 - w_1||z_2 - w_2|}{|z_1 - w_2||z_2 - w_1|}.$$

Hence

$$v \le 32^{\,2(K-1)} \max(u^K, u^{-K}),$$

whence

$$\log v \le K|\log u| + b(K) \le K a_D(z_1, z_2) + b(K)$$

by Lemma 10.4.1. Thus

$$a_{f(D)}(f(z_1), f(z_2)) \le K a_D(z_1, z_2) + b(K).$$

□

COROLLARY 10.4.7. *If D is a domain in $\overline{\mathbf{R}^2}$ and if there exists a K-quasiconformal self-mapping f of $\overline{\mathbf{R}^2}$ which maps D conformally onto a disk, then*

$$h_D(z_1, z_2) \le K\, a_D(z_1, z_2) + b(K). \tag{10.4.8}$$

PROOF. By (3.3.8) and Theorem 10.4.4,

$$\begin{aligned} h_D(z_1, z_2) &= h_{f(D)}(f(z_1), f(z_2)) \\ &= a_{f(D)}(f(z_1), f(z_2)) \le K\, a_D(z_1, z_2) + b(K). \end{aligned}$$

□

THEOREM 10.4.9. *If $D \subset \overline{\mathbf{R}^2}$ is a K-quasidisk, then*

$$h_D(z_1, z_2) \le K^2\, a_D(z_1, z_2) + b(K^2) \tag{10.4.10}$$

for $z_1, z_2 \in D$.

PROOF. By hypothesis, there exists a K-quasiconformal self-mapping f of $\overline{\mathbf{R}^2}$ which maps D onto a disk B. The existence theorem for the Beltrami equation implies there exists a K-quasiconformal self-mapping $g : B \to B$ such that $g \circ f$ is conformal in D. See, for example, Ahlfors [**6**] or Lehto-Virtanen [**117**]. Reflection in $\partial\mathbf{B}$ extends g to a K-quasiconformal self-mapping of $\overline{\mathbf{R}^2}$. Then $\tilde{f} = g \circ f$ is K^2-quasiconformal and we can apply Corollary 10.4.7 to obtain (10.4.10). □

The bound in (10.4.10) can be replaced by an inequality of the form

$$h_D(z_1, z_2) \le c(K)\, a_D(z_1, z_2) \tag{10.4.11}$$

where $c(K) \to 1$ as $K \to 1$. To show this, we must first establish (10.4.11) for the case where $a_D(z_1, z_2)$ is small. The proof for this fact depends in an essential way on the geometry of ∂D since, as noted earlier, $a_D(z_1, z_2) = 0$ whenever $z_1, z_2 \in D$ are symmetric in a circle C which contains ∂D.

Our argument makes use of the following sharp variant of the three-point condition for a quasicircle. See Corollary 2.24 in Gehring-Hag [**58**].

COROLLARY 10.4.12. *If z_1, z_2, z_3, z_4 is an ordered quadruple of points on a K-quasicircle C and if*

$$\max(|z_1 - z_0|, |z_3 - z_0|) \le a \le b \le \min(|z_2 - z_0|, |z_4 - z_0|)$$

for some point $z_0 \in \mathbf{R}^2$, then

$$b/a \le \lambda(K)^{1/2}, \tag{10.4.13}$$

where

$$\lambda(K) = \left(\frac{1}{4}\, e^{\pi K/2} - e^{-\pi K/2}\right)^2 + \delta(K), \qquad 0 < \delta(K) < e^{-\pi K}.$$

THEOREM 10.4.14. *If $D \subset \mathbf{R}^2$ is a K-quasidisk, then*

$$h_D(z_1, z_2) \le \lambda(K)\, a_D(z_1, z_2)$$

for each pair of points $z_1, z_2 \in D$ with

$$a_D(z_1, z_2) \le \lambda(K)^{-1/2}. \tag{10.4.15}$$

PROOF. Set $m = \lambda(K)^{1/2} \ge 1$ and choose distinct $z_1, z_2 \in D$ which satisfy (10.4.15). By performing a preliminary Möbius transformation we may assume that $z_1 = 0$, $z_2 = \infty$ and that

$$r - 1 = \inf_{z \notin D} |z| > 0, \qquad r + 1 = \sup_{z \notin D} |z| < \infty.$$

Then

$$\mathbf{R}^2 \setminus \overline{D} \subset \{z : r - 1 < |z| < r + 1\}$$

and

$$2/r < \log\left(\frac{r+1}{r-1}\right) = a_D(0, \infty) \le 1/m \tag{10.4.16}$$

by (10.4.15). Hence $m < r/2$. Next

$$C = \partial D \subset \{z : r - 1 \le |z| \le r + 1\}.$$

and by performing a preliminary rotation, we may assume that

$$\{\arg(z) : z \in \mathbf{R}^2\} = [-\phi, \phi]$$

where $\phi > 0$.

We shall show that $\phi \le \pi/6$. For this fix $0 < \theta < \min(\phi, \pi/2)$ and choose $w_1, w_2, w_3, w_4 \in C$ so that $\arg(w_2) = \theta$, $\arg(w_4) = -\theta$ and so that w_1 and w_3 are positive and separate w_2, w_4 in C. Then w_1, w_2, w_3, w_4 is an ordered quadruple of points on the K-quasicircle C with

$$\max(|w_1 - r|, |w_3 - r|) \le 1.$$

Hence

$$r\sin(\theta) \le \min\left(|w_2 - r|, |w_4 - r|\right) \le m < r/2$$

by Corollary 10.4.12 and $\theta < \pi/6$. Thus $\phi \le \pi/6$ and

$$C \subset \{z : r - 1 \le |z| \le r + 1,\ |\arg(z)| \le \pi/6\}.$$

Let U denote the smallest closed disk with center on $\{z : |z| = r\}$ which contains C; by performing a second preliminary rotation we may assume that

$$U = \{z :\ |z - r| \le b\}.$$

Then $b \le m$, for if $b > 1$, there exists as above an ordered quadruple of points $\tilde{w}_1, \tilde{w}_2, \tilde{w}_3, \tilde{w}_4$ on C such that

$$\max(|\tilde{w}_1 - r|, |\tilde{w}_3 - r|) \le 1 < b = |\tilde{w}_2 - r| = |\tilde{w}_4 - r|$$

and $b \le m$ by Corollary 10.4.12.

Finally, set

$$f(z) = \frac{m}{r - z}.$$

Then $\mathbf{B} \subset f(D)$ and

$$h_D(0, \infty) = h_{f(D)}(m/r, 0) \le h_{\mathbf{B}}(m/r, 0) = \log\left(\frac{r + m}{r - m}\right).$$

Hence if

$$g(t) = \log\left(\frac{t + m}{t - m}\right) - m^2 \log\left(\frac{t + 1}{t - 1}\right),$$

then $g'(t) > 0$ for $r < t < \infty$ and we obtain

$$m^2\, a_D(0, \infty) - h_D(0, \infty) \ge \lim_{s \to \infty} (g(s) - g(r)) = \int_r^\infty g'(t)\, dt > 0.$$

□

COROLLARY 10.4.17. *If $D \subset \overline{\mathbf{R}^2}$ is a K-quasidisk, then*

$$h_D(z_1, z_2) \le c(K)\, a_D(z_1, z_2) \tag{10.4.18}$$

for $z_1, z_2 \in D$ where $c(K) \to 1$ as $K \to 1$.

PROOF. Fix $z_1, z_2 \in D$. Then

$$h_D(z_1, z_2) \le \lambda(K)\, a_D(z_1, z_2)$$

by Theorem 10.4.14 if

$$a_D(z_1, z_2) \le \lambda(K)^{-1/2} \le 1$$

while

$$h_D(z_1, z_2) \le \left(K^2 + \lambda(K)^{1/2}\, b(K^2)\right)\, a_D(z_1, z_2)$$

by Theorem 10.4.9 otherwise. Hence (10.4.18) holds with

$$c(K) = \max\left(\lambda(K), K^2 + \lambda(K)^{1/2}\, b(K^2)\right).$$

Next $\lambda(K) \to 1$ and $b(K) \to 0$, whence $c(K) \to 1$, as $K \to 1$. □

10.5. Apollonian metric and hyperbolic segments

The following consequence of Theorem 10.3.3 completes the second chain of implications in this chapter.

THEOREM 10.5.1. *Suppose that D is simply connected and that*

$$h_D(z_1, z_2) \le b\, a_D(z_1, z_2) \tag{10.5.2}$$

for each $z_1, z_2 \in D$ where b is a constant. Then for each hyperbolic segment γ joining $z_1, z_2 \in D$ and each $z \in \gamma$,

$$\begin{aligned} \text{length}(\gamma) &\le c\, |z_1 - z_2|, \\ \min_{j=1,2} \text{length}(\gamma_j) &\le c\, \text{dist}(z, \partial D) \end{aligned} \tag{10.5.3}$$

where γ_1, γ_2 are the components of $\gamma \setminus \{z\}$ and $c = c(b)$.

PROOF. Lemma 3.3.5 and (10.5.2) imply that

$$h_D(z_1, z_2) \le b\, a_D(z_1, z_2) \le b\, j_D(z_1, z_2)$$

for each $z_1, z_2 \in D$. Hence (10.5.3) follows for each hyperbolic segment γ joining $z_1, z_2 \in D$ and each $z \in \gamma$ from Theorem 10.3.3. □

CHAPTER 11

Fourth series of implications

In this chapter we prove the following statements for a simply connected domain D.

1° A quasidisk is a quasiconformal extension domain.
2° A quasiconformal extension domain is homogeneous with respect to the family $QC(K)$ for a fixed K.
3° A domain homogeneous with respect to the family $QC(K)$ for a fixed K is a quasidisk.
4° A quasidisk has the extremal distance property.
5° A domain with the extremal distance property is a linearly locally connected domain.
6° A Jordan domain is a quadsidisk if and only if it has the harmonic bending property.

These implications are indicated in Figure 11.1.

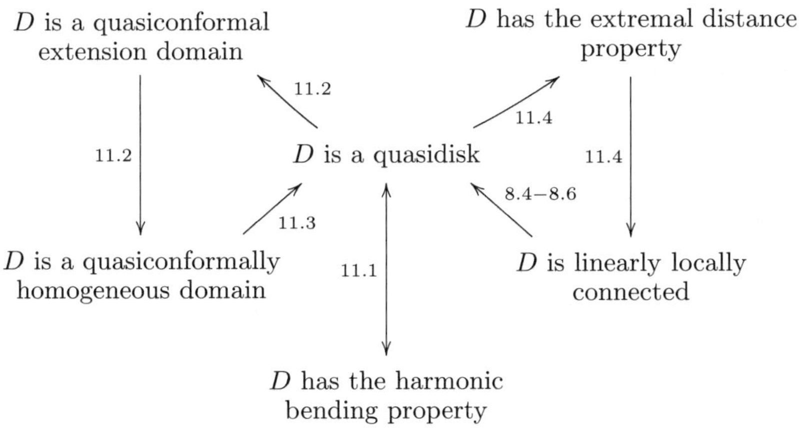

Figure 11.1

11.1. Harmonic bending and quasidisks

We show first that a Jordan domain $D \subset \mathbf{R}^2$ is a quasidisk if and only if it has the harmonic bending property, that is, if each arc $\gamma \subset \partial D$ does not bend too close to $\partial D \setminus \gamma$, the complement of γ in ∂D. See Fernández-Hamilton-Heinonen [**41**].

We begin with the following bound for the harmonic measure of a boundary arc in a Jordan domain. See McMillan [**128**].

LEMMA 11.1.1. *Suppose that D is a Jordan domain in $\mathbf{R}^2$ and that γ is an arc in ∂D. Then*

$$\omega(z_0, \gamma; D) \le \frac{4}{\pi} \arctan \left(\frac{\operatorname{diam}(\gamma)}{\operatorname{dist}(z_0, \gamma)} \right)^{1/2} \tag{11.1.2}$$

for each point $z_0 \in D$.

PROOF. Suppose that $z_0 \in D$, choose $z_1 \in \gamma$ so that

$$|z_0 - z_1| = \operatorname{dist}(z_0, \gamma),$$

and let $d = \operatorname{diam}(\gamma)$. Then by a change of variable we may assume that $d = 1$, $z_0 = -|z_0| < 0$, and $z_1 = 0$. We may also assume that $|z_0| > 1$ since otherwise

$$\omega(z_0, \gamma; D) \le 1 = \frac{4}{\pi} \arctan(1) \le \frac{4}{\pi} \arctan \left(\frac{1}{|z_0|} \right)^{1/2}$$

and we obtain (11.1.2).

Since D is a Jordan domain with $\gamma \subset \partial D \cap \mathbf{B}$, there exists an arc β_1 which joins a point $w_1 \in \partial \mathbf{B}$ to ∞ such that

$$(\beta_1 \setminus \{w_1\}) \cap (D \cup \overline{\mathbf{B}}) = \emptyset.$$

Let D_1 be the domain with $\partial \mathbf{B} \cup \beta_1$ as its boundary and let $\gamma_1 = \partial \mathbf{B} \setminus \{w_1\}$. Then $\omega(z, \gamma; D) - \omega(z, \gamma_1; D_1)$ is harmonic in $D \setminus \overline{\mathbf{B}}$,

$$\omega(z, \gamma; D) = 0 \le \omega(z, \gamma_1; D_1)$$

in $\partial(D \setminus \overline{\mathbf{B}})$, and hence

$$\omega(z_0, \gamma; D) \le \omega(z_0, \gamma_1; D_1). \tag{11.1.3}$$

Next let β_2 denote the part of the real axis joining $w_2 = 1$ to ∞, let D_2 be the domain with $\partial \mathbf{B} \cup \beta_2$ as its boundary, and set $\gamma_2 = \partial \mathbf{B} \setminus \{w_2\}$. Then $\omega(z, \gamma_2; D_2)$ is harmonic in D_2 and

$$\omega(z_0, \gamma_1; D_1) \le \omega(z_0, \gamma_2; D_2) \tag{11.1.4}$$

by the solution of the Carleman-Milloux Problem. See, for example, Nevanlinna [**138**, pp. 107-113].

Finally, let

$$f(z) = \frac{1}{2} \left(\sqrt{z} + \frac{1}{\sqrt{z}} \right)$$

so that $f(1) = 1$, $f(-1) = -1$, and

$$f(z_0) = \frac{i}{2} \left(\sqrt{|z_0|} - \frac{1}{\sqrt{|z_0|}} \right) = \frac{i}{2} (\cot(\phi) - \tan(\phi)) = \frac{i}{\tan(2\phi)}$$

where

$$\phi = \arctan \left(\frac{1}{|z_0|} \right)^{1/2}.$$

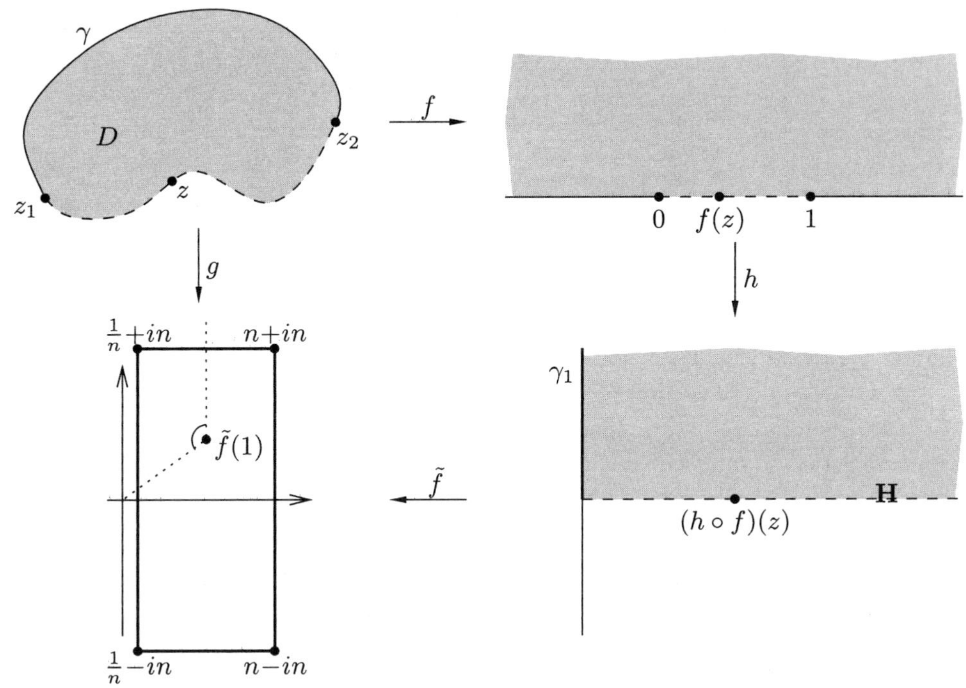

FIGURE 11.2

Hence

$$\omega(z_0, \gamma_2; D_2) = \omega(f(z_0), f(\gamma_2); f(D_2)) \tag{11.1.5}$$
$$= \frac{2}{\pi} \arctan\left(\frac{1}{|f(z_0)|}\right) = \frac{4}{\pi} \arctan\left(\frac{1}{|z_0|}\right)^{1/2}$$

and (11.1.2) follows from (11.1.3) and (11.1.4). □

The above argument shows that there exists an example for which inequality in Lemma 11.1.1 is sharp whenever $\operatorname{diam}(\gamma) \leq \operatorname{dist}(z_0, \gamma)$ and that the simpler bound

$$\omega(z_0, \gamma; D) \leq \frac{4}{\pi} \left(\frac{\operatorname{diam}(\gamma)}{\operatorname{dist}(z_0, \gamma)}\right)^{\frac{1}{2}} \tag{11.1.6}$$

holds for each arc γ in ∂D and each point $z_0 \in D$.

We turn now to the proof of the following characterization for quasidisks in terms of harmonic bending.

THEOREM 3.9.3 (Fernández-Hamilton-Heinonen [**41**]). *A Jordan domain D is a K-quasidisk if and only if it has the harmonic bending property with constant c, where K and c depend only on each other.*

PROOF. For necessity assume that $C = \partial D$ is a quasicircle. Let γ be a closed arc of C with endpoints z_1, z_2 and consider $z \in C \setminus \gamma$. Then there exists a K-quasiconformal map f of $\overline{\mathbf{R}}^2$ which maps C onto $\overline{\mathbf{R}}$ and $C \setminus \gamma$ onto $(0,1)$. In order to use Definition 3.9.2 of harmonic bending we consider a preliminary conformal

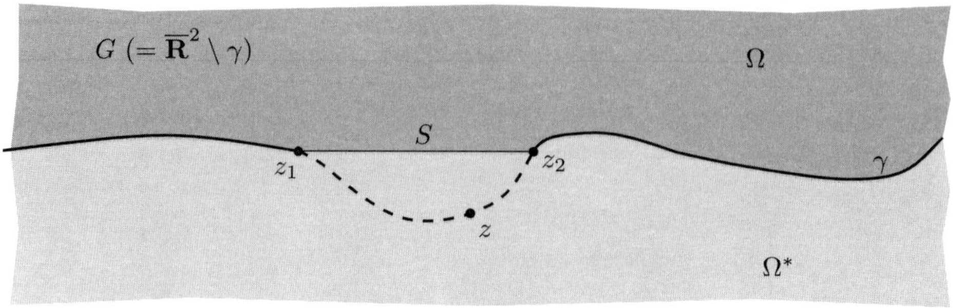

FIGURE 11.3

map h of $\overline{\mathbf{R}}^2 \setminus f(\gamma)$ onto $\mathbf{H}$. Without loss of generality we let $f(\gamma)$ correspond to γ_1, the positive half of the imaginary axis. Using an additional Möbius transformation if necessary, we may also assume that $h(f(z)) = 1$.

Finally, there exists a K-quasiconformal map $\tilde{f}$ of $\mathbf{H}$ fixing $0, i, \infty$ such that the composition $g = \tilde{f} \circ h \circ f$ is conformal. This is a consequence of the measurable Riemann mapping theorem, Theorem 1.1.11, and the composition rules for the complex dilatation. Hence by symmetry D has the harmonic bending property if

$$\omega(g(z), \gamma_1; \mathbf{H}) \le c \tag{11.1.7}$$

where $c = c(K)$. But this follows from the fact that the collection of K-quasiconformal maps of $\mathbf{H}$ fixing three points on the boundary is a normal family. Hence there must exist an $n \in \mathbf{N}$, $n = n(K)$, such that $\tilde{f}(1)$ is in the rectangle $[\frac{1}{n}, n] \times [-n, n]$ and

$$\omega(g(z), \gamma_1; \mathbf{H}) \le \omega(\tfrac{1}{n} + in, \gamma_1; \mathbf{H}) = c.$$

For sufficiency we assume that $\infty \in C = \partial D$ (which we may do since the harmonic bending property is Möbius invariant). Then C is a K-quasicircle if there exists a constant K such that when z_1, z, and z_2 are three finite points on C in that order, then

$$|z_1 - z| < K|z_1 - z_2|.$$

Moreover, we may assume that the open segment S connecting z_1 and z_2 does not intersect C. We normalize the situation so that $z_1 = 0$ and $z_2 = 1$ and denote $C \setminus C(z_1, z_2)$ by γ and $\overline{\mathbf{R}}^2 \setminus \gamma$ by G as in the definition of harmonic bending. Finally, Ω and Ω^* denote the the components of $\overline{\mathbf{R}}^2 \setminus (\gamma \cup S)$, and we assume without loss of generality that $z \in \Omega^*$.

Since D has the harmonic bending property with constant c, we know that

$$\omega(z, \partial\Omega^* \cap \partial G; G) \le c.$$

Here the left-hand side is the solution of the Dirichlet problem whose boundary data is 1 on the Ω^*-side of γ and 0 on the Ω-side, evaluated at z. Likewise

$$\omega(z, \partial\Omega \cap \partial G; G) \le c.$$

Then

$$\omega(z, \partial\Omega \cap \partial G; G) \le \omega(z, S; \Omega^*)$$

since by monotonicity

$$\omega(z, \partial\Omega^* \cap \partial G; G) \geq \omega(z, \gamma; \Omega^*).$$

But by (11.1.6)

$$\omega(z, S; \Omega^*) \leq \frac{4}{\pi\sqrt{d(z,S)}}$$

and

$$1 - \frac{4}{\pi\sqrt{d(z,S)}} \leq \omega(z, \partial\Omega^* \cap \partial G; G) \leq c < 1$$

so that

$$|z_1 - z| - 1 \leq d(z,S) \leq \left(\frac{4}{\pi(1-c)}\right)^2 + 1.$$

This means that

$$\frac{|z_1 - z|}{|z_2 - z_1|} = |z_1 - z| \leq \left(\frac{4}{\pi(1-c)}\right)^2 + 2.$$

□

11.2. Quasidisks and quasiconformal extension domains

THEOREM 11.2.1. *If D is a K-quasidisk, then any K'-quasiconformal self-mapping f of D has a cK'-quasiconformal extension as a self-mapping of $\overline{\mathbf{R}}^2$, where c only depends on K.*

PROOF. We may assume that $D = g(\mathbf{B})$ for a K-quasiconformal mapping g of $\overline{\mathbf{R}}^2$. Then the mapping $h = g^{-1} \circ f \circ g$ is a K^2K'-quasiconformal self-mapping of $\mathbf{B}$ and thus has a K^2K'-quasiconformal extension to $\overline{\mathbf{R}}^2$ by Example 5.3.2. Then the mapping $g \circ h \circ g^{-1}$ is a K^4K'-quasiconformal extension of g, so we may take $c = K^4$. □

Thus quasidisks are quasiconformal extension domains.

Next we prove that quasiconformal extension domains are homogeneous with respect to the family of mapping QC.

THEOREM 11.2.2. *If a simply connected domain D is a quasiconformal extension domain with constant c, then it is homogeneous with respect to $QC(c)$.*

PROOF. For this, fix $z_1, z_2 \in D$, let f be a conformal mapping of D onto $\mathbf{B}$, and let $w_j = f(z_j)$ for $j = 1, 2$. Then there exists a Möbius transformation $g : \mathbf{B} \to \mathbf{B}$ such that $g(w_1) = w_2$ and hence $h = f^{-1} \circ g \circ f$ is a conformal self-mapping of D which maps z_1 onto z_2. Then since D is a quasiconformal extension domain, h has a c-quasiconformal extension to $\overline{\mathbf{R}}^2$, and thus D is homogeneous. □

11.3. Homogeneity and quasidisks

We will give a proof due to Sarvas [**153**] for the case when the homogeneity is with respect to quasiconformal maps of the domain. The argument is a clever application of the three-point condition for quasidisks. See Theorem 2.2.5.

THEOREM 11.3.1. *A Jordan domain D is a quasidisk if it is homogeneous with respect to the family $QC(K)$ for some fixed K.*

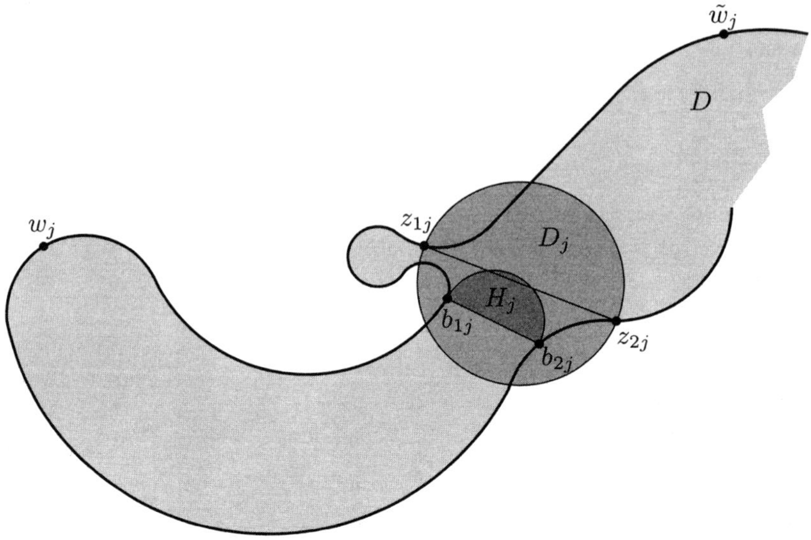

FIGURE 11.4

PROOF. We may assume without loss of generality that D is bounded. Suppose next that D is homogeneous with respect to $QC(K)$ for some $1 \leq K < \infty$ but that D is not a quasidisk. We will show that this leads to a contradiction.

Since D is not a quasidisk, there exists a sequence of ordered triples of points (z_{1j}, w_j, z_{2j}) in ∂D which violates the three-point condition, i.e., such that

$$\lim_{j\to\infty} \frac{|w_j - z_{2j}|}{|z_{1j} - z_{2j}|} = \infty. \tag{11.3.2}$$

Let γ_j and γ_j' be the components of $\partial D \setminus \{z_{1j}, z_{2j}\}$ labeled so that $w_j \in \gamma_j$. Since D is bounded, (11.3.2) implies that

$$\lim_{j\to\infty} (z_{1j} - z_{2j}) = 0 \tag{11.3.3}$$

and hence that

$$\lim_{j\to\infty} \operatorname{diam}(\gamma_j) = 0, \qquad \lim_{j\to\infty} \operatorname{diam}(\gamma_j') = \operatorname{diam}(\gamma) > 0.$$

Thus we may assume without loss of generality that

$$\operatorname{diam}(\gamma_j) \leq \operatorname{diam}(\gamma_j').$$

We may also assume that each open segment (z_{1j}, z_{2j}) and at least one of the two open half-disks with (z_{1j}, z_{2j}) as a common boundary, which we denote by H_j, does not intersect ∂D. To see this, let

$$|w_j - z_{2j}| > 2|z_{1j} - z_{2j}|$$

and choose $w_j' \in \gamma_j'$ so that

$$2\,|w_j' - z_{2j}| \geq |w_j - z_{2j}| > 2\,|z_{1j} - z_{2j}| = 4r_j.$$

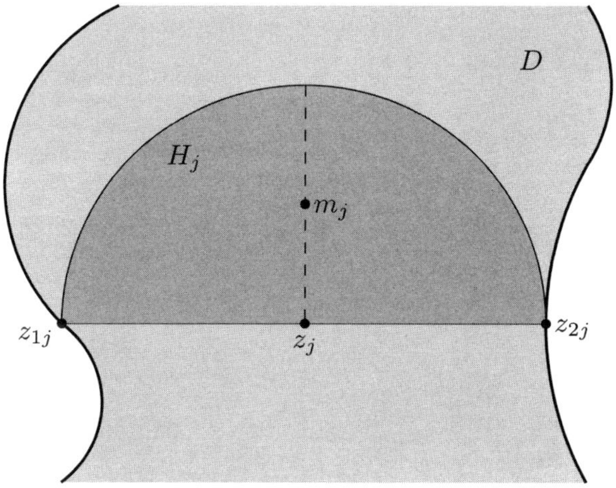

FIGURE 11.5

Next let β_j and β_j' be the components of $\partial D \setminus \{w_j, w_j'\}$ and choose $\zeta_{1j} \in \beta_j$ and $\zeta_{2j} \in \beta_j'$ so that

$$|\zeta_{1j} - \zeta_{2j}| = \operatorname{dist}(\beta_j \cap \overline{D}_j, \beta_j' \cap \overline{D}_j)$$

where

$$D_j = \mathbf{B}\Big(\frac{z_{1j} + z_{2j}}{2}, r_j\Big), \qquad r_j = \frac{|z_{1j} - z_{2j}|}{2}.$$

See Figure 11.4. Then

$$(\zeta_{1j}, \zeta_{2j}) \cap \partial D = \emptyset$$

and w_j and w_j' are in different components of $\partial D \setminus \{\zeta_{1j}, \zeta_{2j}\}$. In addition

$$\frac{|w_j - \zeta_{2j}|}{|\zeta_{1j} - \zeta_{2j}|} \to \infty \quad \text{and} \quad \frac{|w_j' - \zeta_{2j}|}{|\zeta_{1j} - \zeta_{2j}|} \to \infty.$$

If we replace z_{1j} and z_{2j} by ζ_{1j} and ζ_{2j}, and w_j by w_j' if necessary, we obtain the sequence of open segments (z_{1j}, z_{2j}) described above.

By passing to a subsequence of the triples (z_{1j}, w_j, z_{2j}) we can divide the rest of the proof for Theorem 11.3.1 into the following two cases.

1° For each j the open half-disk H_j and the segment (z_{1j}, z_{2j}) lie in D.

2° Every segment (z_{1j}, z_{2j}) lies in the complement D^*.

Proof for case 1

Fix $a \in D$ and let w_j' be a point in γ_j' such that

$$2\,|w_j' - z_{2j}| \geq |w_j - z_{2j}|\,.$$

Let H_j be the open half-disk described above, set $z_j = \frac{1}{2}(z_{1j} + z_{2j})$, and let m_j be the midpoint of the segment joining z_j to the midpoint of the circular boundary of H_j. See Figure 11.5 for an illustration.

Next choose $f_j \in QC(K)$ such that

$$f_j(D) = D, \qquad f_j(a) = m_j,$$

let $L_j : \overline{\mathbf{R}}^2 \to \overline{\mathbf{R}}^2$ be an affine mapping so that

$$L_j(H_j) = \mathbf{B} \cap \mathbf{H} = H, \qquad L_j(z_{1j}) = -1, \qquad L_j(z_{2j}) = 1,$$

and let $g_j = L_j \circ f_j$ for $j = 1, 2, \ldots$. Then

$$g_j(a) = L_j(m_j) = i/2$$

for all j and from the three-point condition we infer that $|L_j(w_j)|$ and $|L_j(w'_j)|$ tend to infinity, since

$$\frac{|w_j - z_{2j}|}{|z_{1j} - z_{2j}|} = \frac{|L_j(w_j) - 1|}{2} \to \infty, \qquad \frac{|w'_j - z_{2j}|}{|z_{1j} - z_{2j}|} = \frac{|L_j(w'_j) - 1|}{2} \to \infty.$$

Because $z_{1j}, z_{2j} \in \partial D$, the sequence of maps g_j omits the points $-1, 1$ and is normal in D. Hence $\{g_j\}$ has a subsequence which converges uniformly on compact subsets to $g_0 : D \to \overline{D}$, where g_0 is either a K-quasiconformal mapping or a constant. See Lehto-Virtanen [**117**]. Since $H \subset g_j(D)$ for every j, the constant must lie in $\overline{g_0(D)} \setminus H$. But we know that $g_j(a) = i/2$ so this is impossible. Hence g_0 is a K-quasiconformal mapping.

By passing to a subsequence once more, we conclude that the sequence $\{g_j\}$ converges uniformly to a K-quasiconformal mapping $g : \overline{\mathbf{R}}^2 \to \overline{\mathbf{R}}^2$ where $g|_D = g_0$. Moreover, g is injective and since the inverse mappings g_j^{-1} converge uniformly to g^{-1}, we may assume that

$$\lim_{j\to\infty} f_j^{-1}(z_{1j}) = z_1 \in \partial D, \qquad \lim_{j\to\infty} f_j^{-1}(z_{2j}) = z_2 \in \partial D,$$
$$\lim_{j\to\infty} f_j^{-1}(w_j) = w \in \partial D, \qquad \lim_{j\to\infty} f_j^{-1}(w'_j) = w' \in \partial D.$$

By the construction of g, $g(z_1) = -1$, $g(z_2) = 1$, and $g(w) = g(w') = \infty$. But this implies that $w = w'$ and that w is different from the distinct points z_1, z_2. This is impossible.

Proof for case 2

We may assume that

$$|w_j - z_{2j}| = \sup\{|w - z_{2j}| : w \in \overline{\gamma_j}\} > |z_{1j} - z_{2j}|$$

for all j. Then since $\gamma_j \subset \overline{\mathbf{B}}(z_{2j}, |w_j - z_{2j}|)$ while $\operatorname{dist}(w_j, \gamma'_j) > 0$, there is an $r_j > 0$ such that

$$(11.3.4) \qquad B_j = \mathbf{B}(c_j, r_j\,|w_j - z_{2j}|) \subset D$$

with $c_j = w_j + r_j(w_j - z_{2j})$.

Next by passing to a subsequence we assume that $r_j = r > 0$ for all j. To see this, let r_j be maximal in (11.3.4) and choose

$$w'_j \in (\partial D \setminus \{w_j\}) \cap \overline{B}_j.$$

If

$$\lim_{j\to\infty} r_j = 0, \qquad \lim_{j\to\infty} |w'_j - w_j| = 0,$$

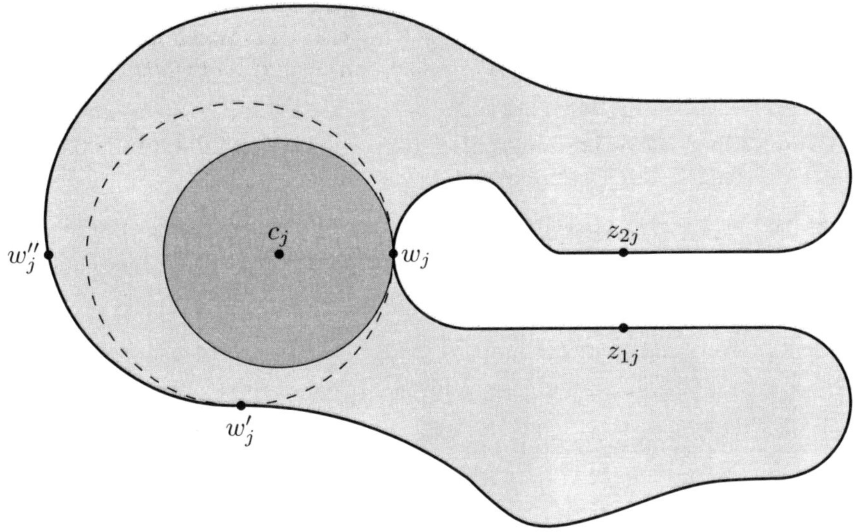

FIGURE 11.6

then the points w_j', w_j, z_{2j} form triples which reduce the situation to case 1, since by (11.3.2)

$$\frac{\min(\operatorname{diam}\beta_j, \operatorname{diam}\beta_j')}{|w_j' - w_j|} \geq \frac{\min\{|w_j - z_{2j}|, |w_j - z_{1j}|\}}{|w_j' - w_j|} \geq \frac{|w_j - z_{2j}| - |w_j - z_{1j}|}{2r_j\,|w_j - z_{2j}|} \to \infty$$

where β_j and β_j' denote the components of $\partial D \setminus \{w_j, w_j'\}$.

The situation $r_j = r > 0$ for all j in (11.3.4) can also be treated much the same way as in case 1. Fix $a \in D$ and let $f_j \in QC(K)$ satisfy $f_j(D) = D$ and $f_j(a) = c_j$. Next let $L_j : \overline{\mathbf{R}}^2 \to \overline{\mathbf{R}}^2$ be an affine map so that $L_j(B_j) = \mathbf{B}$ and $L(w_j) = 1$. Again the composed mappings $g_j|_D$ omit 1 and ∞. Moreover, since $g_j(a) = 0 \in g_j(D)$ for all j, we conclude as before that, by passing to subsequences if necessary, the sequences (g_j) and (g_j^{-1}) converge uniformly to K-quasiconformal mappings g and g^{-1}, respectively.

Now let w_j'' be the last point on the ray $\{w_j + t(w_j - z_{2j}) : t \geq 0\}$ which meets ∂D. We may suppose that

$$\lim_{j\to\infty} f_j^{-1}(z_{1j}) = z_1 \in \partial D, \quad \lim_{j\to\infty} f_j^{-1}(z_{2j}) = z_2 \in \partial D,$$
$$\lim_{j\to\infty} f_j^{-1}(w_j) = w \in \partial D, \quad \lim_{j\to\infty} f_j^{-1}(w_j'') = w'' \in \partial D.$$

Here

$$\left|g_j(f_j^{-1}(z_{1j})) - g_j(f_j^{-1}(z_{2j}))\right| = |L_j(z_{1j}) - L_j(z_{2j})| = \frac{1}{r}\,|z_{1j} - z_{2j}|,$$

which tends to 0 by (11.3.3). Therefore $z_1 = z_2$. Furthermore,

$$g_j(f_j^{-1}(z_{2j})) = 1 + \frac{1}{r}, \quad g_j(f_j^{-1}(w_j)) = 1, \quad g_j(f_j^{-1}(w_j'')) = -s$$

with $s > 0$, so that z_1, w, and w'' are distinct points. We get a contradiction as before. □

11.4. Extremal distance domains

That quasidisks have the extremal distance property follows immediately from the following remark and Example 3.11.1.

REMARK 11.4.1. If a domain D has the extremal distance property with constant c and if $f : \overline{\mathbf{R}}^2 \to \overline{\mathbf{R}}^2$ is K-quasiconformal, then $f(D)$ has the extremal distance property with constant K^2c.

PROOF. Given continua C_1 and C_2 in $\overline{\mathbf{R}}^2$ and in D, respectively, then $f(\Gamma)$ and $f(\Gamma_D)$ are the families of curves joining $f(C_1)$ and $f(C_2)$ in $\overline{\mathbf{R}}^2$ and in $f(D)$, respectively, and

$$\operatorname{mod}(f(\Gamma)) \leq K \operatorname{mod}(\Gamma) \leq Kc \operatorname{mod}(\Gamma_D) \leq K^2 c \operatorname{mod}(f(\Gamma_D)).$$

□

The converse statement that a simply connected domain D with the extremal distance property is a quasidisk will be proved by showing that extremal distance domains must be linearly locally connected. The result then follows from the first series of implications in Chapter 8.

Our proof of linear local connectivity depends on the following lower bound for the modulus of a curve family.

LEMMA 11.4.2. *If Γ is the family of curves which join disjoint continua C_1 and C_2 in $\overline{\mathbf{R}}^2$ and if $z_1, w_1 \in C_1$ and $z_2, w_2 \in C_2$, then*

$$\operatorname{mod}(\Gamma) \geq \frac{\pi}{\log 4\sqrt{r}} \qquad \text{where} \qquad r = \frac{|z_1 - z_2||w_1 - w_2|}{|z_1 - w_1||z_2 - w_2|}.$$

PROOF. By performing a preliminary Möbius transformation we need only consider the case where $z_1 = 0, w_1 = 1, |z_2| = r \geq 1$, and $w_2 = \infty$. Then

$$\operatorname{mod}(\Gamma) = \frac{2\pi}{\operatorname{mod} R_T(r)},$$

where $R_T(r)$ is the ring domain bounded by the intervals $[0, 1]$ and $[r, \infty]$, and we obtain

$$\operatorname{mod} R_T(r) \leq \log 16r = 2 \log 4\sqrt{r}.$$

See, for example, pages 173, 55, 61 in Lehto-Virtanen [**117**]. □

Lemma 11.4.2 yields the following lower bound for the modulus of a curve family Γ which joins continua C_1 and C_2 in $\overline{\mathbf{R}}^2$.

LEMMA 11.4.3. *Suppose that C_1 and C_2 are disjoint continua in $\overline{\mathbf{R}}^2$ and that*

$$\min_{j=1,2} \operatorname{diam}(C_j) \geq a \operatorname{dist}(C_1, C_2),$$

where $a > 0$. If Γ is the family of curves which join C_1 and C_2 in $\overline{\mathbf{R}}^2$, then

$$\operatorname{mod}(\Gamma) \geq \frac{\pi a}{2a + 1}.$$

PROOF. Choose $z_1 \in C_1$ and $z_2 \in C_2$ so that

$$|z_1 - z_2| = \operatorname{dist}(C_1, C_2).$$

Then by hypothesis there exists a point $w_j \in C_j$ such that

$$|z_j - w_j| \geq \frac{1}{2}\operatorname{diam}(C_j) \geq \frac{a}{2}\operatorname{dist}(C_1, C_2) = \frac{a}{2}|z_1 - z_2|$$

for $j = 1, 2$. By relabeling if necessary we may also assume that

$$|z_1 - w_1| \leq |z_2 - w_2|.$$

Let $f : \overline{\mathbf{R}}^2 \to \overline{\mathbf{R}}^2$ be a Möbius transformation for which $f(w_2) = \infty$. Then

$$\begin{aligned}\frac{|f(z_1) - f(z_2)|}{|f(z_1) - f(w_1)|} &= \frac{|z_1 - z_2||w_1 - w_2|}{|z_1 - w_1||z_2 - w_2|} \leq \frac{2}{a}\frac{|w_1 - w_2|}{|z_2 - w_2|} \\ &\leq \frac{2}{a}\frac{|z_1 - w_1| + |z_1 - z_2| + |z_2 - w_2|}{|z_2 - w_2|} \leq 4\frac{a+1}{a^2} = t,\end{aligned}$$

and it is not difficult to show that

$$\log 4\sqrt{t} = \log\left(\frac{8\sqrt{a+1}}{a}\right) < \frac{2a+1}{a}.$$

We conclude that

$$\operatorname{mod}(\Gamma) = \operatorname{mod}(f(\Gamma)) \geq \frac{\pi}{\log 4\sqrt{t}} \geq \frac{\pi a}{2a+1}$$

by Lemma 11.4.2. □

A set $E \subset \overline{\mathbf{R}}^2$ is said to be *a-convex* where $1 \leq a < \infty$ if each pair of points $z_1, z_2 \in E \setminus \{\infty\}$ can be joined by a curve $\gamma \subset E$ such that

$$\operatorname{length}(\gamma) \leq a|z_1 - z_2|.$$

If $E \subset \mathbf{R}^2$, then E is 1-convex if and only if E is convex in the usual sense.

We show next that domains with the extremal distance property are a-convex for some constant a.

LEMMA 11.4.4 (Gehring-Martio [**65**]). *If $D \subset \overline{\mathbf{R}}^2$ has the extremal distance property with constant c, then D is a-convex where*

$$a = (5/3)\exp(48\,c). \tag{11.4.5}$$

PROOF. Fix $z_1, z_2 \in \mathbf{R}^2$, let $r = |z_1 - z_2|$, and let C be a curve joining z_1 and z_2 in D. Next for $j = 1, 2$ let C_j denote the component of

$$C \cap \overline{\mathbf{B}}(z_j, r/4)$$

that contains z_j and let Γ_D and Γ denote the families of curves joining C_1 and C_2 in D and $\overline{\mathbf{R}}^2$, respectively. Then

$$\min_{j=1,2}\operatorname{diam}(C_j) \geq \frac{r}{4} \geq \frac{\operatorname{dist}(C_1, C_2)}{4}$$

and hence by Lemma 11.4.3 and (3.11.4)

$$\operatorname{mod}(\Gamma_D) \geq \frac{\operatorname{mod}(\Gamma)}{c} \geq \frac{b_0}{c}, \tag{11.4.6}$$

where $b_0 = \pi/6$.

Now let Γ_1 be the family of curves in Γ_D which lie in $\mathbf{B}(z_2, s)$ where

$$s = \frac{r}{4}\exp\Big(\frac{4\pi c}{b_0}\Big) = r\, b_1, \tag{11.4.7}$$

and let $\Gamma_2 = \Gamma_D \setminus \Gamma_1$. If each curve $\gamma \in \Gamma_1$ has length $l(\gamma) \geq l$, then

$$\rho(z) = \begin{cases} l^{-1} & \text{if } |z - z_2| \leq s, \\ 0 & \text{otherwise} \end{cases}$$

is in $\mathrm{adm}(\Gamma_1)$ and hence

$$\mathrm{mod}(\Gamma_1) \leq \frac{\pi s^2}{l^2}. \tag{11.4.8}$$

Next each curve $\gamma \in \Gamma_2$ joins the circles $|z - z_2| = r/4$ and $|z - z_2| = s$. Thus

$$\rho(z) = \begin{cases} (\log(4s/r)\,|z - z_2|)^{-1} & \text{if } r/4 \leq |z - z_2| \leq s, \\ 0 & \text{otherwise} \end{cases}$$

is in $\mathrm{adm}(\Gamma_2)$ and

$$\mathrm{mod}(\Gamma_2) \leq \frac{2\pi}{\log(4s/r)}. \tag{11.4.9}$$

Inequalities (11.4.6), (11.4.8), and(11.4.9) with (11.4.7) then imply that

$$\frac{b_0}{c} \leq \mathrm{mod}(\Gamma_D) \leq \mathrm{mod}(\Gamma_1) + \mathrm{mod}(\Gamma_2) \leq \frac{\pi s^2}{l^2} + \frac{b_0}{2c}$$

and we obtain

$$l \leq \sqrt{\frac{2\pi c}{b_0}}\, s = \sqrt{\frac{2\pi c}{b_0}}\,\frac{r}{4}\,\exp(\frac{4\pi c}{b_0}) \leq r\,\exp(b_1 c),$$

where $b_1 = 8\pi/b_0 = 48$.

Set $b_2 = \exp(b_1 c)$. Then there is a rectifiable curve $\gamma_0 \in \Gamma_D$ with

$$\mathrm{length}(\gamma_0) \leq b_2 r = b_2|z_1 - z_2|$$

and endpoints w_1 and $\tilde{w}_1$ in C such that

$$|z_1 - w_1| \leq |z_1 - z_2|/4, \qquad |z_2 - \tilde{w}_1| \leq |z_1 - z_2|/4.$$

Next set $r_1 = |z_1 - w_1|$ and let C_1 and C_2 denote the components of $C \cap \overline{\mathbf{B}}(z_1, r_1/4)$ and $\gamma_0 \cap \overline{\mathbf{B}}(w_1, r_1/4)$ which contain z_1 and w_1, respectively. Then

$$\min_{j=1,2} \mathrm{diam}(C_j) \geq \frac{r_1}{4} \geq \frac{1}{4}\mathrm{dist}(C_1, C_2)$$

and arguing as above we obtain a curve γ_1 in D with

$$\mathrm{length}(\gamma_1) \leq b_2\frac{|z_1 - z_2|}{4^2}$$

such that γ_1 joins γ_0 to a point $w_2 \in C$ with

$$|z_1 - w_2| \leq \frac{1}{4^2}|z_1 - z_2|.$$

An analogous procedure yields a subcurve $\tilde{\gamma}_1$ in D with

$$\mathrm{length}(\tilde{\gamma}_1) \leq b_2\frac{|z_1 - z_2|}{4^2}$$

such that $\tilde{\gamma}_1$ joins γ_0 to a point $\tilde{w}_2 \in C$ with

$$|z_2 - \tilde{w}_2| \leq \frac{1}{4^2}|z_1 - z_2|.$$

Continuing in this way we obtain two sequences of curves $\gamma_1, \gamma_2, \ldots, \gamma_n, \ldots$ and $\tilde{\gamma}_1, \tilde{\gamma}_2, \ldots, \tilde{\gamma}_n, \ldots$ whose union together with γ_0 contains a rectfiable curve γ joining z_1 and z_2 in D with

$$\begin{aligned}
\operatorname{length}(\gamma) &\leq \operatorname{length}(\gamma_0) + \sum_{n=1}^{\infty} \operatorname{length}(\gamma_n) + \sum_{n=1}^{\infty} \operatorname{length}(\tilde{\gamma}_n) \\
&\leq b_2\big(|z_1 - z_2| + \sum_{n=1}^{\infty} \frac{|z_1 - z_2|}{4^n} + \sum_{n=1}^{\infty} \frac{|z_1 - z_2|}{4^n}\big) \\
&= b|z_1 - z_2|,
\end{aligned}$$

where $b = (5/3)\exp(48c)$. □

We turn now to the main result in this section.

THEOREM 11.4.10. *If a domain $D \subset \overline{\mathbf{R}}^2$ has the extremal distance property with constant c, then D is linearly locally connected with constant b where*

$$b \leq 4\exp(48c). \tag{11.4.11}$$

PROOF. Fix $z_0 \in \mathbf{R}^2$ and $r > 0$. Then D is a-convex with

$$a \leq (5/3)\exp(48c)$$

by Lemma 11.4.4. Suppose that $z_1, z_2 \in D \cap \mathbf{B}(z_0, r)$. Then z_1 and z_2 can be joined by a curve $\gamma \subset D$ with

$$\operatorname{length}(\gamma) \leq a|z_1 - z_2| \leq 2ar,$$

whence

$$|z - z_0| \leq |z - z_1| + |z_1 - z_0| \leq (2a + 1)r = br$$

for $z \in \gamma$. Thus $\gamma \subset \mathbf{B}(z_0, br)$ and $D \cap \mathbf{B}(z_0, r)$ lies in a component of $D \cap \mathbf{B}(z_0, br)$.

Finally, if $f : \overline{\mathbf{R}}^2 \to \overline{\mathbf{R}}^2$ is a Möbius transformation, then $f(D)$ has the extremal distance property with constant c by Remark 11.4.1 and is therefore a-convex by Lemma 11.4.4. Hence $f(D) \cap \mathbf{B}(z_0, r)$ lies in a component of $f(D) \cap \mathbf{B}(z_0, br)$ by the argument given above, and D is b-linearly locally connected by Remark 2.4.2. □

Bibliography

[1] S. B. Agard, *Distortion theorems for quasiconformal mappings.* Ann Acad. Sci. Fenn. **413** (1968) 1-11.

[2] S. B. Agard and F. W. Gehring, *Angles and quasiconformal mappings.* Proc. London Math. Soc. **14A** (1965) 1-21.

[3] L. V. Ahlfors, *Complex Analysis.* McGraw-Hill, 1979.

[4] L. V. Ahlfors, *Remarks on the Neumann-Poincaré integral equation.* Pacific J. Math. **2** (1952) 271-280.

[5] L. V. Ahlfors, *Quasiconformal reflections.* Acta Math. **109** (1963) 291-301.

[6] L. V. Ahlfors, *Extension of quasiconformal mappings from two to three dimensions.* Proc. Nat. Acad. Sciences **51** (1964) 768-771.

[7] L. V. Ahlfors, *Lectures on quasiconformal mappings.* 2nd. ed., University Lecture Series **38**, American Mathematical Society (2006).

[8] L. V. Ahlfors, *Conformal invariants.* McGraw-Hill, 1973.

[9] L. V. Ahlfors and L. Bers, *Riemann's mapping theorem for variable metrics.* Ann. of Math. **72** (1960) 385-404.

[10] P. Alestalo, D. A. Herron, and J. Luukkainen, *Ahlfors' three-point property.* Complex Var. Elliptic Equ. **41** (2000) 327-329.

[11] G. D. Anderson, M. K. Vamanamurthy, and M. Vuorinen, *Distortion functions for plane quasiconformal mappings.* Israel J. Math. **62** (1988) 1-16.

[12] G. D. Anderson, M. K. Vamanamurthy, and M. Vuorinen, *Conformal invariants, inequalities, and quasiconformal maps.* John Wiley & Sons, 1997.

[13] K. Astala, *Area distortion of quasiconformal mappings.* Acta Math. **173** (1994) 37-60.

[14] K. Astala and F. W. Gehring, *Injectivity, the BMO norm and the universal Teichmüller space.* J. Anal. Math. **46** (1986) 16-57.

[15] K. Astala, K. Hag, P. Hag, and V. Lappalainen, *Lipschitz classes and the Hardy-Littlewood property.* Monatsh. Math. **115** (1993) 267-279.

[16] K. Astala, T. Iwaniec, and G. Martin, *Elliptic Partial Differential Equations and Quasiconformal Mappings in the Plane.* Princeton University Press, 2008.

[17] A. F. Beardon, *On the Hausdorff dimension of general Cantor sets.* Proc Camb. Phil. Soc. **61** (1965) 679-694.

[18] A. F. Beardon, *The Geometry of Discrete Groups.* Springer-Verlag, 1983.

[19] A. F. Beardon, *Iteration of Rational Functions.* Springer-Verlag, 1991.

[20] A. F. Beardon, *The Apollonian metric of a domain in* $\mathbf{R}^n$. Quasiconformal mappings and analysis, Springer-Verlag, 1998, 91-108.

[21] J. Becker, *Löwnersche Differentialgleichung und quasikonform fortsetzbare schlichte Funktionen.* J. Reine Angew. Math. **255** (1972) 23-43.

[22] J. Becker and C. Pommerenke, *Schlichtheitskriterien und Jordangebiete.* J. reine angew. Math. **354** (1984) 74-94.

[23] A. Beurling and L. V. Ahlfors, *The boundary correspondence under quasiconformal mappings.* Acta Math. **96** (1956) 125-142.

[24] C. J. Bishop, *Bilipschitz homogeneous curves in* $\mathbf{R}^2$ *are quasicircles.* Trans. Amer. Math. Soc. **353** (2002) 2655-2663.

[25] B. Bojarski, *On the Beltrami equation, once again: 54 years later.* Ann. Acad. Sci. Fenn. **35** (2010) 59–73.

[26] M. Bonk, *The support points of the unit ball in Bloch space.* J. Functional Anal. **123** (1994) 318-335.

[27] B. Brechner and T. Erkama, *On topologically and quasiconformally homogeneous continua.* Ann. Acad. Sci. Fenn. **4** (1978/1979) 207-208.

[28] O. J. Broch, *Geometry of John disks.* Ph.D. thesis, NTNU (2005).

[29] O. J. Broch, K. Hag, and S. Junge, *A note on the harmonic measure doubling condition.* Conform. Geom. Dyn. **15** (2011) 1-6.

[30] D. Calvis, *The inner radius of univalence of normal circular triangles and regular polygons.* Complex Var. Elliptic Equ. **4** (1985) 295-304.

[31] L. Carelson and T. W. Gamelin, *Complex Dynamics.* Sringer-Verlag, 1993.

[32] M. Chuaqui and B. G. Osgood, *John domains, quasidisks, and the Nehari class.* J. reine angew. Math. **471** (1996) 77-114.

[33] P. L. Duren, *Univalent Functions.* Springer-Verlag, 1983.

[34] P. L. Duren, *Theory of H^p Spaces.* Dover, 2000.

[35] P. L. Duren, H. S. Shapiro, and A. L. Shields, *Singular measures and domains not of Smirnov type.* Duke Math. J. **33** (1966) 247-254.

[36] T. Erkama, *Group actions and extension problems for maps of balls.* Ann. Acad. Sci. Fenn. **556** (1973) 3-31.

[37] T. Erkama, *Quasiconformally homogeneous curves.* Michigan Math. J. **24** (1977) 157-159.

[38] T. Erkama, *Möbius automorphisms of plane domains.* Ann. Acad. Sci. Fenn. **10** (1985) 155-162.

[39] C. Fefferman, *Characterizations of bounded mean oscillation.* Bull. Amer. Math. Soc. **77** (1971) 587-588.

[40] C. Fefferman and E. M. Stein, *H^p spaces of several variables.* Acta Math. **129** (1972) 137-193.

[41] J. L. Fernández, D. Hamilton, and J. Heinonen, *Harmonic measure and quasicircles.* (Preprint).

[42] J. L. Fernández, J. Heinonen, and O. Martio, *Quasilines and conformal mappings.* J. Anal. Math. **52** (1989) 117-132.

[43] B. B. Flinn, *Hyperbolic convexity and level sets of analytic functions.* Indiana Univ. Math. J. **32** (1983) 831-841.

[44] J. B. Garnett, *Bounded Analytic Functions.* Academic Press, 1981.

[45] J. B. Garnett and D. E. Marshall, *Harmonic Measure.* Cambridge Univ. Press, 2005.

[46] J. B. Garnett, F. W. Gehring, and P. W. Jones, *Conformally invariant length sums.* Indiana Univ. Math. J. **32** (1983) 809-829.

[47] F. W. Gehring, *The definitions and exceptional sets for quasiconformal mappings.* Ann. Acad. Sci. Fenn. **281** (1960) 3-28.

[48] F. W. Gehring, *Quasiconformal mappings of slit domains in three space.* J. Math. Mech. **18** (1969) 689-703.

[49] F. W. Gehring, *Univalent functions and the Schwarzian derivative.* Comment. Math. Helv. **52** (1977) 561-572.

[50] F. W. Gehring, *Injectivity of local quasi-isometries.* Comment. Math. Helv. **57** (1982) 202-220.

[51] F. W. Gehring, *Characteristic properties of quasidisks.* Les Presses de l'Université de Montréal, 1982, 1-97.

[52] F. W. Gehring, *Extension of quasiisometric embeddings of Jordan curves.* Complex Var. Elliptic Equ. **5** (1986) 245-263.

[53] F. W. Gehring, *Uniform domains and the ubiquitous quasidisk.* Jber. Deutsche Math. Verein. **89** (1987) 88-103.

[54] F. W. Gehring, *Characterizations of quasidisks.* Banach Center Publications **48** (1999) 11-41.

[55] F. W. Gehring and K. Hag, *Remarks on uniform and quasiconformal extension domains.* Complex Var. Elliptic Equ. **9** (1987) 175-188.

[56] F. W. Gehring and K. Hag, *Hyperbolic geometry and disks.* J. Comp. Appl. Math. **104** (1999) 275-284.

[57] F. W. Gehring and K. Hag, *A bound for hyperbolic distance in a quasidisk.* Computational Methods and Function Theory 1997, World Scientific Publishing Co., 1999, 233-240.

[58] F. W. Gehring and K. Hag, *The Apollonian metric and quasiconformal mappings.* Contemp. Math. **256** (2000) 143-163.

[59] F. W. Gehring and K. Hag, *Reflections on reflections in quasidisks*, Report. Univ. Jyväskylä **83** (2001) 81-90.

[60] F. W. Gehring and K. Hag, *Sewing homeomorphisms and quasidisks.* Comput. Methods Funct. Theory **3** (2003) 143-150.

[61] F. W. Gehring, K. Hag, and O. Martio, *Quasihyperbolic geodesics in John domains.* Math. Scand. **65** (1989) 75-92.

[62] F. W. Gehring and W. K. Hayman, *An inequality in the theory of conformal mapping.* J. Math. Pures Appl. **41** (1962) 353-361.

[63] F. W. Gehring and O. Lehto, *On the total differentiability of functions of a complex variable.* Ann. Acad. Sci. Fenn. **272** (1959) 3-9.

[64] F. W. Gehring and O. Martio, *Quasidisks and the Hardy-Littlewood property.* Complex Var. Elliptic Equ. **2** (1983) 67-78.

[65] F. W. Gehring and O. Martio, *Quasiextremal distance domains and extension of quasiconformal mappings.* J. Anal. Math. **45** (1985) 181-206.

[66] F. W. Gehring and O. Martio, *Lipschitz classes and quasiconformal mappings.* Ann. Acad. Sci. Fenn. **10** (1985) 203-219.

[67] F. W. Gehring and B. G. Osgood, *Uniform domains and the quasi-hyperbolic metric.* J. Anal. Math. **36** (1979) 50-74.

[68] F. W. Gehring and B. P. Palka, *Quasiconformally homogeneous domains.* J. Anal. Math. **30** (1976) 172-199.

[69] F. W. Gehring and Ch. Pommerenke, *On the Nehari univalence criterion and quasicircles.* Comment. Math. Helv. **59** (1984) 226-242.

[70] F. W. Gehring and J. Väisälä, *Hausdorff dimension and quasiconformal mappings.* J. London Math. Soc. **6** (1973) 504-512.

[71] J. Gevirtz, *Injectivity of quasi-isometric mappings of balls.* Proc. Amer. Math. Soc. **85** (1982) 345-349.

[72] M. Ghamsari, *Sobolev and quasiconformal extension domains.* Proc. Amer. Math. Soc. **119** (1993) 1179-1188.

[73] M. Ghamsari and D. A. Herron, *Bi-Lipschitz homogeneous Jordan curves.* Trans. Amer. Math. Soc. 351 (1999) 3197-3216.

[74] M. Ghamsari, R. Näkki, and J. Väisälä, *John disks and extension of maps.* Monatsh. Math. **117** (1994) 63-94.

[75] V. M. Gol′dstein and S. K. Vodop′janov, *Prolongement des fonctions de classe L_1^p et applications quasi conformes.* C. R. Acad. Sc. Paris **290** (1980) 453-456.

[76] V. M. Gol′dstein, T. G. Latfullin, and S. K. Vodop′janov, *Criteria for extension of functions of the class L_2^1 from unbounded plane domains.* Siberian Math. J. **20** (1979) 298-301.

[77] K. Hag, *What is a disk?*, Banach Center Publications **48** (1999) 43-53.

[78] W. K. Hayman, *Multivalent Functions.* Cambridge Univ. Press, 1958.

[79] W. K. Hayman and J.-M. G. Wu, *Level sets of univalent functions.* Comment. Math. Helv. **56** (1981) 366-403.

[80] J. Heinonen, *Lectures on Analysis on Metric Spaces.* Springer-Verlag, 2001.

[81] J. Heinonen and P. Koskela, *Definitions of quasiconformality.* Invent. Math. **120** (1995) 61-79.

[82] D. A. Herron, *The geometry of uniform, quasicircle, and circle domains.* Ann. Acad. Sci. Fenn. **12** (1987) 217-228.

[83] D. A. Herron, *John domains and the quasihyperbolic metric.* Complex Var. Elliptic Equ. **39** (1988) 327-334.

[84] D. A. Herron and P. Koskela, *Quasiextremal distance domains and conformal mappings onto circle domains.* Complex Var. Elliptic Equ. **15** (1990) 167-179.

[85] J. Hersch, *Contribution à la théorie des fonctions pseudo-analytiques.* Comm. Math. Helv. **30** (1956) 1-18.

[86] G. A. Hjelle, *A simply connected, homogeneous domain that is not a quasidisk.* Ann. Acad. Sci. Fenn. **30** (2005) 135-143.

[87] S. Jaenisch, *Length distortion of curves under conformal mappings.* Michigan Math. J. **15** (1968) 121-128.

[88] D. S. Jerison and C. E. Kenig, *Hardy spaces, A_∞, and singular integrals on chord-arc domains.* Math. Scand. **50** (1982) 221-247.

[89] D. S. Jerison and C. E. Kenig, *Boundary behaviour of harmonic functions in non-tangentially accessible domains.* Adv. Math. **46** (1982) 80-147.

[90] F. John, *Rotation and strain.* Comm. Pure Appl. Math. **14** (1961) 391-413.

[91] F. John, *On quasi-isometric mappings, I.* Comm. Pure Appl. Math. **21** (1968) 77-110.

[92] F. John, *On quasi-isometric mappings, II.* Comm. Pure Appl. Math. **22** (1969) 265-278.

[93] F. John and L. Nirenberg, *On functions of bounded mean oscillation.* Comm. Pure Appl. Math. **14** (1961) 415-426.

[94] P. W. Jones, *Extension theorems for BMO.* Indiana Univ. Math. J. **29** (1980) 41-66.

[95] P. W. Jones, *Quasiconformal mappings and extendability of functions in Sobolev spaces.* Acta Math. **147** (1981) 71-88.

[96] S. Kallunki and P. Koskela, *Exceptional sets for definition of quasiconformality.* Amer. J. Math. 122 (2000) 735-743.

[97] S. Kallunki and P. Koskela, *Metric definition of μ-homeomorphisms.* Michigan Math. J. **51** (2003) 141-151.

[98] S. Kallunki and O Martio, *ACL homeomorphisms and linear dilatation.* Proc. Amer. Math. Soc. **130** (2002) 1073-1078.

[99] R. Kaufman and J.-M. Wu, *Distances and the Hardy-Littlewood Property.* Complex Var. Elliptic Equ. **4** (1984) 1-5.

[100] J. A. Kelingos, *Boundary correspondence under quasiconformal mappings.* Michigan Math. J. **13** (1966) 235-249.

[101] K. Kim, *Harmonic doubling condition and John disks.* Comm. Korean Math. Soc. **10** (1995) 145-153.

[102] K. Kim and N. Langmeyer, *Harmonic measure and hyperboic distance in John disks.* Math. Scand. **83** (1998), 283-299.

[103] B. N. Kimel′fel′d, *Homogeneous regions on the conformal sphere.* Mat. Zametki **8** (1970) 321-328. (Russian)

[104] W. Kraus, *Über den Zusammenhang einiger Charakteristiken eines einfach zusammenhängend Bereiches mit der Kreisabbildung.* Mitt. Math. Sem. Giessen **21** (1932) 1-28.

[105] J. G. Krzyż, *Quasicircles and harmonic measure.* Ann. Acad. Sci. Fenn. **12** (1987) 19-24.

[106] J. G. Krzyż, *Harmonic analysis and boundary correspondence under quasiconformal mappings.* Ann. Acad. Sci. Fenn. **14** (1989) 225-242.

[107] J. G. Krzyż, *Quasisymmetric functions and quasihomographies.* Ann. Univ. M. Curie-Skłodowska **47** (1993) 90-95.

[108] R. Kühnau, *Möglichst konforme Spiegelung an einer Jordankurve.* Jber. Deut. Math.-Verein **90** (1988) 90-109.

[109] R. Kühnau, *Möglichst konforme Spiegelung an einer Jordanbogen auf der Zahlenkugel.* Complex Analysis, Birkhäuser Verlag, 1988, 139-156.

[110] N. Langmeyer, *The quasihyperbolic metric, growth, and John domains.* Ann. Acad. Sci. Fenn. **23** (1998) 205-224.

[111] V. Lappalainen, Lip_h*-Extension domains.* Ann. Acad. Sci. Fenn. Dissertationes **56** (1985) 56 pp.

[112] M. Lehtinen, *On the inner radius of univalency for non-circular domains.* Ann. Acad. Sci. Fenn. **5** (1980) 45-47.

[113] M. Lehtinen, *Angles and the inner radius of univalency.* Ann. Acad. Sci. Fenn. **11** (1986) 151-165.

[114] O. Lehto, *Domain constants associated with Schwarzian derivative.* Comment Math. Helv. **52** (1977) 603-610.

[115] O. Lehto, *Remarks on Nehari's theorem about the Schwarzian derivative and schlicht functions.* J. Anal. Math. **36** (1979) 184-190.

[116] O. Lehto, *Univalent Functions and Teichmüller Spaces.* Springer-Verlag, 1987.

[117] O. Lehto and K. I. Virtanen, *Quasiconformal mappings in the plane.* Springer-Verlag, 1973.

[118] O. Lehto, K. I. Virtanen, and J. Väisälä, *Contributions to the distortion theory of quasiconformal mappings.* Ann. Acad. Sci. Fenn. **273** (1959) 3-13.

[119] X. Liu and D. Minda, *Distortion theorems for Bloch functions.* Trans. Amer. Math. Soc. **333** (1992) 325-338.

[120] P. MacManus, *Bi-Lipschitz extensions in the plane.* J. Anal. Math. **66** (1995) 85-115.

[121] P. MacManus, R. Näkki, and B. Palka, *Quasiconformally homogeneous compacta in the complex plane.* Michigan Math. J. **45** (1998) 227-241.

[122] P. MacManus, R. Näkki, and B. Palka, *Quasiconformally bi-homogeneous compacta in the complex plane.* Proc. London Math. Soc. (3) **78** (1999) 215-240.
[123] O. Martio and J. Sarvas, *Injectivity theorems in plane and space.* Ann. Acad. Sci. Fenn. **4** (1978-1979) 383-401.
[124] B. Maskit, *On Klein's combination theorem.* Trans. Amer. Math. Soc. **120** (1965) 499-509.
[125] B. Maskit, *On boundaries of Teichmüller spaces and on Kleinian groups: II.* Ann. of Math. **91** (1970) 607-639.
[126] B. Maskit, *Kleinian Groups.* Springer-Verlag, 1987.
[127] P. Mattila, *Geometry of Sets and Measures in Euclidean Spaces.* Cambridge Univ. Press, 1995.
[128] J. E. McMillan, *Boundary behavior under conformal mapping.* Proc. N. R. L. Conf. Classical Function Theory, Washington, 1970, 59-76.
[129] D. Menchoff, *Sur une généralisation d'un théorème de M. H. Bohr.* Sbornik. Mat. **44** (1937) 339-354.
[130] J. Miller, *Sector reflections in the plane.* Ann. Acad. Sci. Fenn. **30** (2005) 219-225.
[131] L. Miller-Van Wieren, *Univalence criteria on classes of rectangles and equiangular hexagons.* Ann. Acad. Sci. Fenn. **22** (1997) 407-424.
[132] D. Minda, *The Schwarzian derivative and univalence criteria.* Contemp. Math. **38** (1985) 43-52.
[133] C. B. Morrey, *On the solution of quasilinear elliptic partial differential equations.* Trans Amer. Math. Soc. **43** (1938) 126-166.
[134] J. Moser, *On Harnack's theorem for elliptic differential equations.* Comm. Pure Appl. Math. **14** (1961) 577-591.
[135] R. Näkki and B. P. Palka, *Quasiconformal circles and Lipschitz classes.* Comm. Math. Helv. **55** (1980) 485-498.
[136] R. Näkki and J. Väisälä, *John disks.* Exp. Math. **9** (1991) 3-43.
[137] Z. Nehari, *The Schwarzian derivative and schlicht functions.* Bull. Amer. Math. Soc. **55** (1949) 545-551.
[138] R. Nevanlinna, *Über die Methode der sukzessiven Approximationen.* Ann. Acad. Sci. Fenn. **291** (1960) 1-10.
[139] R. Nevanlinna, *Analytic Functions.* Springer-Verlag, 1970.
[140] M. H. A. Newman, *The topology of plane sets of points.* Cambridge Univ. Press, 1954.
[141] B. G. Osgood, *Some properties of f''/f' and the Poincaré metric.* Indiana Univ. Math. J. **31** (1982) 449-461.
[142] K. Øyma, *Harmonic measure and conformal length.* Proc. Amer. Math. Soc. **115** (1992) 687-689.
[143] K. Øyma, *The Hayman-Wu constant.* Proc. Amer. Math. Soc. **119** (1993) 337-338.
[144] A. Pfluger, *Über die Konstruktion Riemannscher Flächen durch Verheftung.* J. Indian Math. Soc. **24** (1960) 401-412.
[145] C. Pommerenke, *Boundary Behaviour of Conformal Maps.* Springer-Verlag, 1992.
[146] H. M. Reimann, *Functions of bounded mean oscillation and quasiconformal mappings.* Comment. Math. Helv. **49** (1974) 260-276.
[147] H. M. Reimann and T. Rychener, *Funktionen beschränkter mittlerer Oszillation.* Lecture Notes in Math. **487**, Springer-Verlag, 1975.
[148] S. Rickman, *Characterization of quasiconformal arcs.* Ann. Acad. Sci. Fenn. **395** (1966) 3-30.
[149] S. Rickman, *Extension over quasiconformally equivalent curves.* Ann. Acad. Sci. Fenn. **436** (1969) 3-12.
[150] S. Rickman, *Quasiconformally equivalent curves.* Duke Math. J. **36** (1969) 387-400.
[151] S. Rohde, *Quasicircles modulo bilipschitz maps.* Rev. Mat. Iberoamericana **17** (2001) 643-659.
[152] S. Rohde, *On the theorem of Hayman and Wu.* Proc. Amer. Math. Soc. **130** (2002) 387-394.
[153] J. Sarvas, *Boundary of a homogeneous Jordan domain.* Ann. Acad. Sci. Fenn. **10** (1985) 511-514.
[154] G. Springer, *Fredholm eigenvalues and quasiconformal mapping.* Acta Math. **111** (1964) 121-142.
[155] D. Stowe, *Injectivity and the pre-Schwarzian derivative.* Michigan Math. J. **40** (1998) 537-546.

[156] D. Sullivan, *On the ergodic theory at infinity of an arbitrary discrete group of hyperbolic motions.* Annals Math. Studies **97**, Princeton Univ. Press, 1981.

[157] O. Teichmüller, *Extremale quasikonforme Abbildungen und quadratische Differentiale.* Abh. Preuss. Akad. Wiss. **22** (1940) 1-197.

[158] O. Teichmüller, *Ein Verschiebungssatz der quasikonformen Abbildung.* Deutsche Math. 7 (1944) 336-343.

[159] M. Tienari, *Fortsetzung einer quasikonformen Abbildung über einen Jordanbogen.* Ann. Acad. Sci. Fenn. A I **321** (1962).

[160] P. Tukia, *The planar Schönflies theorem for lipschitz maps.* Ann. Acad. Sci. Fenn. **5** (1980) 49-72.

[161] P. Tukia, *On two-dimensional quasiconformal groups.* Ann. Acad. Sci. Fenn. **5** (1980) 73-78.

[162] P. Tukia, *Extension of quasisymmetric and Lipschitz embeddings of the real line into the plane.* Ann. Acad. Sci. Fenn. **6** (1981) 89-94.

[163] J. Väisälä, *On quasiconformal mappings of a ball.* Ann. Acad. Sci. Fenn. **304** (1961) 3-7.

[164] J. Väisälä, *Quasimöbius maps.* J. Anal. Math. **44** (1984/85) 218-234.

[165] M. F. Walker, *Linearly locally connected sets and quasiconformal mappings.* Ann. Acad. Sci. Fenn. **11** (1986) 77-86.

[166] R. L. Wilder, *Topology of manifolds.* Colloquium Publications **32**, Amer. Math. Soc., 1949.

[167] K-J. Wirths, *Über holomorphe Funktionen, die einer Wachstumsbeschränkung unterliegen.* Arch. Math. **30** (1978) 606-612.

[168] S. Yang, *QED domains and NED sets in* $\overline{\mathbf{R}}^n$. Trans. Amer. Math. Soc. **334** (1992) 97-120.

[169] S. Yang, *Extremal distance and quasiconformal reflection constants of domains in* $\overline{\mathbf{R}}^n$. J. Anal. Math. **62** (1994) 1-28.

Index